Praise for *On Eating Meat*

Hugh Mackay: 'Compelling, illuminating and often confronting, *On Eating Meat* is a brilliant blend of a gastronome's passion with forensic research into the sources of the meat we eat. Matthew Evans brings his unflinching honesty – and a farmer's hands-on experience – to the question of how to be an ethical carnivore.'

Richard Glover: 'Intellectually thrilling – a book that challenges both vegans and carnivores in the battle for a new ethics of eating. This book will leave you surprised, engrossed and sometimes shocked – whatever your food choices.'

Anton Vikstrom, Good Life Permaculture: 'Matthew Evans fearlessly investigates where our food comes from and the hidden impacts of our industrial food system. If you eat meat, read this book.'

Professor Andy Lowe, Director of Agrifood and Wine, University of Adelaide: 'Insightful, well-researched and highly readable, Matthew Evans' new book *On Eating Meat* presents an honest and challenging assessment of the livestock industry and the ethical and environmental issues surrounding our consumption of meat... this book equips the reader with the knowledge to get beyond the entrenched opinions of its topic area, and allows us to decide whether and what type of meat we wish to consume, and with what consequences for the future.'

Alex Elliott-Howery, Cornersmith: 'This is the most important food book I've read in years. Not just for meat lovers or vegans, it should be read by anyone who eats food. When I finished this book I felt informed, connected and empowered to make better decisions about how I shop, cook and eat.'

Published in 2019 by Murdoch Books, an imprint of Allen & Unwin

Murdoch Books Australia
83 Alexander Street, Crows Nest NSW 2065
Phone: +61 (0)2 8425 0100
murdochbooks.com.au
info@murdochbooks.com.au

Murdoch Books UK
Ormond House, 26–27 Boswell Street,
London WC1N 3JZ
Phone: +44 (0) 20 8785 5995
murdochbooks.co.uk
info@murdochbooks.co.uk

For corporate orders and custom publishing contact our business development team at salesenquiries@murdochbooks.com.au

ISBN 978 1 76063 769 9 Australia
ISBN 978 1 91163 221 4 UK

Tower Hamlets Libraries	
91000008069613	
Askews & Holts	
641.36	
THISBO	TH19-2745/87

Cover design by Design By Committee

Printed and bound in Australia by
10 9 8 7 6 5 4 3 2 1

ON EATING MEAT

The truth about its production
and the ethics of eating it

MATTHEW EVANS

murdoch books
Sydney | London

Contents

Smells Like Money

'It's, um, a bit hard on the nose,' I say, trying to be polite.

'Smells like money,' says the feedlot owner. The chicken farmer. The pig farm manager.

And that, at its heart, is the difference between them and me. I smell ammonia. Something that hits you high and hard in the nostrils and causes an almost involuntary gag. I smell an extraordinary amount of manure, rotting in the sun, or the shed, or the straw. A smell so bad that I have to throw my clothes away after the visit. A smell so toxic that the locals forbid the workers from entering the local café after work until they've changed clothes – and their partners refuse to allow work clothes in the house. It's the smell of intensive animal confinement, one that can lead to ammonia levels in pig sheds to be over four times the recommended 25 parts per million set by Safe Work Australia.

The saying, 'Smells like money', is the difference between the owners of these farms and me. I smell toxins. They smell profit.

Intensive animal farming is the way we largely do things in Australia, and indeed in many countries around the world. It's the way the supermarket $5 roast chicken reaches your shopping basket. How an $8 a kilo ham makes its way to your festive table each Christmas. Why you can enjoy consistently tender beef, despite the cattle being run in feedlots, completely lacking in grass. It's a long way from the idyllic image consumers have of animals grazing contentedly on lush green grass. And it's a long, long way from my little farm in rural Tasmania, where what matters to me is stewardship of the land, the instinctual freedoms of the animals in my care, and an openness to consumers about what we do.

I have many conflictions about my meat eating. Trained as a chef, I later became a food writer and then restaurant critic, more interested in the pedigree of the chefs than the origins of their ingredients. The ethics of food production came secondary to the need to eat a lot (I was very active, and skinny, when young). Any interest in the morality of what we consume also came after the desire to eat food that tasted great.

Yes, for a while after leaving school, like many I dabbled in vegetarianism, vaguely concerned with how the animals I ate were reared, but more interested in the quality and cost of the meat I had been able to afford. And like most who dabble in such diets, I didn't stick to it once the pay cheques started to roll in, and the quality of the meat I could afford improved.

Yet niggles remained in my mind: rumours of intensive farms, of animals denied the ability to express their basic instincts; suggestions that farming of animals wasn't as ideal as we'd all presumed.

Later, wondering why some food tasted better, I ventured onto more farms, and discovered that not all animals are reared the same way. Driven by pursuit of quality, I started to glimpse the ugly truth about what is done in our name on farms across the nation. And eventually, thinking it would be the ultimate way to actually know and trust what you eat, I started farming myself.

A decade later, our family is now growing pretty much everything that hits the table when it comes to fruit, vegetables, dairy and meat. In the process, I've discovered how little I once knew, and how much there is still to learn. About the land, about animals. About those who grow food for the rest of the nation to consume. In the process I've made a documentary on the origins of our meat, *For the Love of Meat*, for SBS television, and been to some of the more intensive farms across the land. And what I've discovered is that I don't want to live on a farm that smells like money.

Let's be clear. I am no angel. Animals have died because of my negligence. Disease has caused unnecessary suffering at my hands. When it comes to the livestock in my care, I am no sage, and am not always as good as I believe I should be. I fail my own standards again and again, when it gets down to the reality of things I do on my farm.

But what of these people who I met on my travels, who rear animals on behalf of the meat eaters of my beloved nation? Who talk about the pungent, nose-wrinkling, brain-stifling smell of the manure of their farms as *smelling like money*? Do they fail the standards of those who trust them to breed, rear and dispatch the animals in their care – the animals they look after on behalf of meat eaters around the country? Do they fail, as a group, to meet *your* expectations, day on day, year on year? Are they accountable to you, if you're a meat eater, for the treatment of the animal, and the environment, as you'd like them to be for the price you are paying?

This is what I endeavoured to find out when filming *For the Love of Meat* in 2015 and 2016. I spent a lot of time visiting butchers and abattoirs, and feedlots and chicken farms. I wanted to see behind the locked gates of Australia's most intensive farms to find out how animals are reared in our communal name. And what I found was, all too often ... a locked gate. The farms I did get into were probably industry best practice, and not everything I saw I liked. But it's not me they need to impress, but rather the meat-eating consumers who buy their products. Surely if you are proud of what you do, on behalf of some of the biggest meat eaters in the world, you'd be open to a visit by a meat-eating journalist?

Want to see behind the doors of the 45,000 sow piggery called Rivalea based near Corowa in New South Wales? Well, their Singaporean shareholders don't think you should. They put shareholders' interests above those who actually consume their products, as – sadly – you'd legally expect of a public company.

Want to talk to Australian Pork Limited, an organisation funded (in part) by the surcharge put on the slaughter of every pig I take to the cutting shop from my little farm? Fine, but don't bring a camera. And don't ask them to a round table conversation that may involve their critics. Because this is an industry that is afraid of scrutiny. Afraid of transparency.

What exactly are they afraid of? That perhaps, just perhaps, what they do in the name of meat eaters nationwide won't stand up to public scrutiny? We had a team of researchers, and made repeated requests for interviews, for visits, to see what happens every day to produce meat in Australia. We were polite, we asked nicely. We didn't want to break the law, or break the trust of the farms we wanted to visit, by jumping the fence and filming illegally. And we were repeatedly refused.

We saw some farms, in fact quite a few, but many others wouldn't let us past the front gate. Is it that most industrial farmers don't want you to *know* how they produce food at the price we demand?

They don't want you to *see* just what chemicals they use, what techniques they employ, what happens beyond the storybook farm you have in your head?

While we don't have an open, fair and accountable farming system, what we do have in Australia is a love of meat. We're a nation that eats just about as much meat as any other, trading places with the United States as the world's biggest meat eaters every couple of years. Australians consume, on average, each year: 111.4 kilograms of meat, including 49 kilograms of chicken, 26 kilograms of pork, about 26 kilograms of beef or veal, and 9 kilograms of lamb, with some mutton, duck and turkey adding to the total. Chicken overtook beef as the most popular meat in Australia in 2006.

The global average meat consumption is just 34.1 kilograms per annum. In 2017, Australians ate four times more beef and veal than the global average, and six times more lamb and mutton (lumped together as one type of meat) than the rest of the world, according to Meat & Livestock Australia's 2018 state of the industry report.

And yet, genuinely informed conversation about what constitutes a 'moral' carnivore remains elusive. Partly, this is because the debate has been seconded by those who want us to implicitly trust what happens behind those locked farm gates – as well as by those who want us to eat *no* meat at all.

Vegans run Animal Liberation. They run Voiceless. Non meat eaters are in charge of Animals Australia. Drawn to these organisations because of their love of animals (usually the cute furry ones, not the slugs and snails and caterpillars and grasshoppers) and their desire to see livestock meet no harm, vegetarians and vegans are the ones speaking on behalf of animals. And good on them; at least someone is. Their overarching agenda, though, is for us to abstain from eating meat altogether.

And I hear their argument. Eating meat has far-ranging consequences. Consequences for the animals we eat, obviously. For our health. And for the planet. But just telling everyone to stop eating meat is naïve, and destined to fail – because as each nation on earth becomes wealthier, its people eat more meat. Moderating meat eating is possible; cutting it out altogether is a pipe dream. And it's the *increase* in meat consumption – both here in Australia as a whole, and globally – that is the problem. Yes, we could and should eat less meat in this nation. Yes, we could survive on no meat or animal products. Those are valid arguments, but the discussion about where, how and what animals we rear for food must be had by those who *choose* to eat meat, not only by those who choose not to eat it.

And yet, that discussion is hard to have. While it bothers me that those who eat no meat want to tell me what is right or wrong, what irks me even more is that those who rear and breed and control the most intensive farms across the land think we have no right to know what is done in our name. I wonder, do they think they are above public scrutiny?

Meat is a commodity. Those in the business of intensive production see animals as a system that needs to be efficient, controllable, and the end product consistent. They see behaviours as something to be managed. They see an animal that grows, that isn't sick, as a happy animal. I, however, see animals as individuals, whose wants and needs aren't the same as my wants and needs, and not always the same as their brothers' or sisters' wants and needs.

Farming, especially confinement farming and the industrial system, is looking for the holy grail of manageability – namely, consistency. But consistency denies nature. Evolution is defined by variation and difference. To make life, any life, consistent, is to actively work against one of the strongest forces in nature.

My aim is to look at the ethical reality of farming, both in relation to the animals we farm, and the animals that are impacted by all farming. But little prepared me for some of the tales I'd unearth.

Those in the industrial food system often believe they are doing the best by the animals in their care. They aren't, generally, trying to be inhumane or cruel. But how they *view* those animals, whose souls they are responsible for, is different to the views of some scientists and often in conflict with community expectations. My view is that intensive farmers see problems and solutions as reductive – i.e. if there's a problem (e.g. disease), fix it with a drug. If there's a behavioural issue (sows fighting, tail biting, cannibalism), then isolate the animal. Or cut off the part that bites, or the part that is bitten. The most intensive systems in the country work at getting the *animal* to give the intensive *farmers* what they want, without always looking at what the animal needs and wants.

I don't see meat as a commodity. I see it as a privilege. As an indulgence. This, then, is my ode to farming animals and eating animals. It's one small farmer's view on meat, its place in our culture, our diet, our farming system. This is my look at death, in all its forms, be it on our farm or off it, within farming systems and outside it.

I think meat eaters need to confront the reality that something dies in their name, and that they should be comfortable with the way it's done. But I also think non meat eaters need to reconcile the fact that more suffering happens outside the farm gate than inside. As we'll see later, more death can be wrought on animals by the growing of grains and vegetables than the production of livestock for meat.

And we all need to be aware that morality, ethics and personal responsibility are not as clear cut as those on both ends of the vegan–intensive farm spectrum would have us believe.

This book will delve into this idea, and others: what it means to farm, and what repercussions this has on the land, the soil, the animals around us. I've been thinking a lot about meat of late, and having lived in cities, and now in the bush on my own farm, I reckon most of us don't have the full picture when it comes to the consequences, costs and true origins of the food we consume. There's a sliding scale, a continuum, where we all must find our place, the spot we feel comfortable living, when we consider our impact on the world.

This book, hopefully, will help people of all persuasions find a place on the scale that sits well with their beliefs. It's not meant as an end, but as a place to start discussions on what it really means to eat meat, or to not eat meat, and the responsibilities some of us take on when we farm on behalf of others.

CHAPTER 1

The View
from the City

It's strange to look back on what I once thought, and what I once knew – or thought I knew. It's discomforting to realise that as a kid brought up in the suburbs, a bloke who trained as a chef in traditional restaurants, and then worked most of his professional life around food in three of Australia's biggest cities, I was desperately out of touch with reality.

Like most city folk, I had a distorted view of farming and our impact on the land. I bushwalked. I had friends whose families had farms. I enjoyed going to the country. But those folk who talked about crutching sheep, and gutting cows, and who owned guns 'for work' – those people were alien to me: a strange breed of people who smelled like oilskin, wore elastic-sided boots and could skin a rabbit as quick as look at it. They knew of blood and dirt and things most grim, and little of it did they openly share with namby-pambies like me.

The mob I hung around with were mostly city people, with the usual romanticised view of 'the bush'. In Australia, The Bush (and let's give it the capitalisation it seems to demand) is a place of hardship, big skies, mateship and sadness. A place of Henry Lawson and Banjo Paterson. The Bush is hardwired into our national psyche, from childhood readings of 'The Loaded Dog' to the glamour of *The Man From Snowy River* on the big screen. Despite being a highly urbanised country, clinging to the fringes of the continent and far away from the arid, red heart, we see ourselves as boundary riders under starlit skies, wearing big hats under the baking sun; a country riding on the sheep's back, hardened to the cow's hide and inconsistent rains. In short, we have taken our poets' version of The Outback and The Bush, added in some modern twists, like helicopter mustering of cattle on million-acre stations, and created a vaster, drier, but no more real version of Old MacDonald's Farm in our heads than its European equivalent. What we've done is simply taken a Beatrix Potter book and added a small dose of drought and myxomatosis, but conveniently ignored what actually happens when humans take over a landscape.

So what *does* happen? Even organic farming, according to at least one scientist I've spoken to, is an alien environment plonked into our ecosystem. What we do to grow food is ruin the homes and shelters of virtually all other species that once relied on the same parcel of land. It matters not whether we grow wheat, or soy, or beef, or pork, if it's any kind of intensive agriculture. Where we actually propagate and till land, we alter the entire complexity of life, both within the soil and above it. When we fell trees to house ourselves or dam a creek to provide water for our gardens, something else suffers. Even the agricultural practices of Aboriginal Australians altered the landscape, through firestick farming and more intricate processes of harvest and rest. The management of species selection for food can only come with a consequence. To imagine otherwise is to go against the fundamentals of not only science, but of common sense.

But common sense doesn't matter when you live in town and lamb comes from a butcher, or on a polystyrene tray wrapped in plastic from a supermarket, as it does more and more these days. It doesn't matter when apple juice arrives in a plastic container, and peas are found in the freezer and not in the pod. Common sense means that food comes at a price, being money and only money – and it's only those 1 per cent of the population who are actually involved in producing food at the primary level who are aware of the other costs. And real costs they are: social, environmental, physical and emotional.

In the city, we don't see death: we see meat, or vegetables. We don't see ecosystems, or nuance. We can feel self-righteousness, or indignation, or shame, though. We can use those emotions to set our own moral compass – but also to misunderstand the world around us. We feel the horror of the deaths of sheep on live export ships, yet also shut out the reality of how the food *we* eat was produced. It came from a nice woman at the farmer's market, after all. Or the cheery butcher down the street.

Perhaps it's a result of this divide between city and country. Perhaps it's the fault of farmers trying to protect us from things that might irk us or make us uncomfortable. But what happens is that we *hide* farming practices, when we should explain. We treat consumers with disrespect, not trusting them with the information on how food is really produced, be it plant or animal. We, as a society, shield our children from death, pretending animals go to a 'better place' or doggy heaven. Our family has been chided for showing our son the last moments of a cow's life – as if humans are somehow separate to this cycle of life and death that is self-evident throughout the rest of nature. As though we are somehow immune. Or without stain.

As a community, we Australians shelter our children, and ourselves, from the grim reality of what is done in our name. Then we complain when we don't like the vision that is thrust upon us by activists or on the evening news. We've cleansed and purged our city lives of the dirty business of how things are actually made, built, grown or farmed. What makes it wrong is that we then marginalise, and sometimes demonise, those who have to do those things for us.

Your home is probably built on farmland. That land was probably once home to an entire ecosystem that was carved and felled and torched to produce a monoculture. Your clothes come from somewhere, are made somehow. Your phone has a cost, not just in terms of money, or human lives in a merciless factory, but of lives of the animals affected for all the mines, wells, crops, and transport that are needed to produce it – and the tip space needed to dispose of it. You might love animals so much that you carry your dog in your handbag, or post funny videos of your cat on the internet. Yet animals die every day so you can eat, clothe and house yourself.

From a city perspective, it's easy to criticise farmers. They, after all, do things in everybody's name, so we all have a vested interest. *How dare they lock pigs in sow stalls! Somebody do something!* And Coles, one of Australia's two largest supermarket chains, does. *How dare they lock chickens in cages to provide eggs!* Hence the rise of the cage-free egg. Those people who shoot ducks in the wetlands each season in Victoria? Barbarians! Neanderthals!

Good versus bad. Right and wrong. No grey areas. It's like we never outgrow our childish view of how food is actually produced. Modern farming doesn't resemble an Alison Lester book, with its idyllic mix of animals and rolling hills. Modern farming is an industry, one that has to make money to survive.

It's easy, when you don't actually have to be involved in primary production, when you don't actually have to create anything from the soil, to pretend you don't notice, or to criticise. And it's easy to distance yourself from the reality that every single thing you do, and your neighbour does, and your society does, has an impact on the living earth. We attack when we should talk. Activists can be all too quick to condemn, not help.

Think of this. The farmland where my farm sits was once home to the Melurkedee people. They lived here for at least 40,000 years by current estimates. Since those first humans arrived, though there's been climate change of sorts, and various other disease and geographical factors, the Melurkedee's impact has also been felt through their manner of firestick farming and foraging. This change was exacerbated when Europeans arrived and cut down a whole lot more trees, grew a whole lot more food, and washed a whole lot more topsoil down the Huon River. A whole ecosystem has been changed, a whole environment – both on land and in the rivers and ocean nearby – altered. To those original humans, the land would be almost unrecognisable now.

We see this same ecological impact with other animals – remove the Tassie devils, or add a feral cat, and environmental consequences inevitably ensue. There are countless stories about the introduction of pest species or the removal of apex predators, and how whole ecosystems have had to adapt.

But with humans, a lot of the time, we tend to think of ourselves as just *here* – that we exist within nature, rather than constantly driving so much of its change. It happens so slowly that perhaps sometimes we don't notice. Other times we might just see it as 'progress', or inevitable. In fishing terms it's known as 'shifting baselines', where what your grandad recalls the fishing was like was what it was *always* like, for *him* – which is different to how I'll think about it when I am a grandfather.

In a nubbin, it boils down to what we experience as young people, and how we think this is how it's always been. But in reality, we have been altering landscapes and ecologies for as long as humans have been around. That's simply nature, evolution, and a complex ecosystem in action. Large marsupials, like the giant kangaroo and giant wombat, fell victim in part to human impact, and now only live on through oral histories and in the fossil record.

So why do we suddenly think that putting up housing developments, growing copious amounts of cotton, digging up the ingredients for cement, or iron or bauxite – why do we think having a human population that is growing exponentially and expects more leisure, greater comforts and a higher standard of living, is going to produce a world that is free from animal suffering, so long as we don't eat meat?

Why does popular culture tend to suggest that only a few animal species are being disadvantaged – the popular ones like pandas and orangutans, and not the entire ecosystem … dung beetles, slugs and aphids included?

Why is it that feeding ourselves is the great big baddie, and that domesticated livestock are the only animals that feel pain, or fear, or that can suffer? What is it about the animals in our care that make them more important than every mouse, rat, locust or starling that will cop it if they stand in the way of vegetable production?

I don't pretend to have the answers to these questions. I just know that a whole lot of other people – those who want us all to stop eating meat, and those who want to rear the cheapest possible meat using questionable farming methods – are those who currently run the agenda.

This book will try to unravel our aversion to what happens in our name, and consider what we can all do to turn things around. It will look at who, really, is responsible for the welfare of all animals, including the animals we directly exploit for human purposes.

CHAPTER 2

The View from
the Feedlot

It's 38°C. I'm standing in the sun, on a windless Queensland day, the air heavy like a blanket. I'm trying to focus my thoughts and energies on the feedlot cattle in front of me.

Today, researching meat, I've come to see the kind of place where most of the cattle raised in northern and central Queensland end up. They end up in rectangular, treeless yards, standing on churned, pooey ground, waiting for god to take them (well, the stock truck to take them to the abattoir). They end up standing with nothing to do but eat, sleep, crap and look for somewhere cooler to be, not unlike cows in a paddock. Unless it's eating, virtually every animal on this feedlot is sheltering as close as it can to the shade cloth that lines only a small portion of each pen. Each pen that *has* some shade cloth, that is.

Each metal-lined enclosure is about a third of a hectare in size – 3000 square metres – and each one has 200 animals standing or lying idly. It adds up to about 15 square metres per beast – so it's not like a stockyard where sheep dogs can clamber over the animals' backs, but it's not as spacious as you might think, especially as most of the time the animals bunch up, huddled near food or drink or shade. The concrete feed troughs, called bunks, border one edge, where the cattle poke their heads out through the metal tubing to get to the grain. On either side are other equally brown poo-lined pens, housing other mobs of previously unfamiliar cattle. The nether reaches of the feedlot, about half the pens, have no shade.

I watch these cattle, whose lives up until they arrived here were ones of ranging free on wide open stations – places that work on stocking densities of tens of hectares per head of cattle, not head of cattle per hectare. Many of these animals would only have seen humans a few times a year, at mustering. Most have only been alive for a year, if that. They've come to these hot, grass-less pens to be 'finished', fattened on grain. They stand on mounds of their own excrement, looking like cattle do. Inscrutable. I can't tell if they're blissful, bored or brain dead.

They're here because in many places the grass has run out for the dry season, or the fodder is so poor that the animals need to put on condition. They're here because this is the end of their lives. A place where they are idle, given high-energy feed, and simply counting down the days until they're meat.

A feedlot is not designed to be a pleasant place. They're just very secure pens, lined with steel tubing, with a sturdy rail over a metre and a half tall, and crisscrossed by gravel laneways. Feedlots are usually well removed from towns and major roads, because of the smell. And the stigma. They're often in semi-arid climes, or places where the rainfall isn't sufficient to cause problems with runoff and wet feet when the rain does come. They're usually close to grain-growing regions, because the real cost of fattening cattle here is feed.

This one has feed troughs down the laneways, where truckies rumble up and down with their loads all day: 3.5 tonnes of food is trucked to the edge of each pen, enough for 200 head a day. The feed is a mix of barley and wheat, with hay, cotton seed (probably genetically modified), almond hulls and more, including vitamins and minerals plus 'supplements' (which the feedlot manager did not specify).

The feedlot rations – mostly roughage, but extremely high in protein and fat compared to grass – are designed to help put on weight, and to increase marbling, the fine lines of fat that lace muscles ... muscles that become the meat humans eat. The animals, once semi-wild and free, are now penned up so their muscles don't do much work. A non-working muscle is more tender to eat – but it also lacks flavour, so the extra fat the animals lay down intramuscularly thanks to their new diet and lack of exercise will go some way to make up for the meat's lack of taste.

This is how 40 per cent of the cattle we eat in Australia end up. Feedlots produce about 80 per cent of the beef that is sold in supermarkets, supplying what is considered the cheaper, commodity end of the market. Standing idly in sun-baked pens, often with animals that are unfamiliar to them, inhaling the deeply unattractive ammonia stench, being given a diet that they've not evolved to eat, and unable to walk around very much. This is how most of the cheap beef we buy is finished, and how the beef industry in our northern climes generally stays afloat. If the animals aren't finished in feedlots here, they're exported to feedlots overseas. The distinct wet and dry seasons, which mean that grass simply struggles to grow for part of the year, along with the heat, impoverished soil and low stocking densities, lend themselves to this kind of grazing and feedlot finishing.

Feedlots are considered, by some, to represent everything that is counter to nature. If you give cattle a choice, however, they will often go to the troughs of grain before they'll head to grass. This much we know. Grain is something cattle rarely see when they range – only when they find seed heads on long grass for part of the year – and they're drawn to it like a kid to fairy floss. Given unlimited free access to only grain, they will gorge themselves on it and possibly die of acidosis (excess acid in the gut), or longer-term health issues. Yet the cattle will virtually always go for grain before grass. This grain, made palatable with molasses and salt, and padded out with plenty of roughage, won't kill these animals in the medium term. Thanks to a gradual introduction of the new feed, the cattle's biome – the bacteria that will ferment the food in the gut – will change, as the animals' bodies adjust to the new diet.

I've been to a feedlot before, but this one is much further north, and far bigger. Like many people, most of what I know about feedlots seems tainted by information from the United States, where some animals barely stand on grass their whole life. Permanently shedded or feedlotted cattle are not really a thing here in Australia, and most animals are only passing through, not destined to live in pens for the long haul; the average is about 10–15 per cent of their lifespan. Some are here for just under two months, some half a year or longer. All have ranged on grass or foraged in rough bush for months before they arrive for the fattening. At the time of my visit, in 2016, about 960,000 cattle nationwide were on feedlots, though this one was more than half empty, able to hold 18,000 at capacity.

A criticism of feedlots is often that cattle lie in their own poo. Any cattle sometimes stand or lie in their own faeces, even when they have a ten-acre grassy paddock to roam in, not just when they're cooped up. But here, in this feedlot, they have no choice. The mound in each pen is a machine-made mountain of manure. The poo is piled up by digger until it becomes unwieldy, and is then carted off site.

Feedlots are humourless places (though maybe it depends on your sense of humour). As I get out of the car next to the pens and nearly pass out from the combination of heat and stench, the owner observes wryly that it 'smells like money'. It's late summer, and the stockyards are far from full. I gag, and wonder what it smells like when it's at capacity.

I ask a lot of questions. Such as the one so many people not in the industry worry about: are the cattle given antibiotics? The staff are happy to show me the sick-bay pens, and their register of all the illnesses and medication. There doesn't seem to be secrecy here, now we're through the gate. As with any mass grouping of animals, including humans, some of the 7500 animals here today are unwell. Some are lame, some losing condition. But only a handful. The art of the workers who ride through the pens each day is to find those not thriving, those on the cusp of illness. They find them, ideally, before anything becomes a problem. I farm, so I know animals get sick. And I know that out on those massive cattle stations, the animals probably get injured and sick far from the homestead and die with no outside care. But I also know that Meat & Livestock Australia research shows 90 per cent of feedlot cattle do get put on antimicrobial agents, ionophores (classed as carboxylic polyether antibiotics), to avoid acidosis when they first go on grain. Surely, such widespread use of medication is a sign of animals that are struggling with their diet?

High acid in the rumen isn't the only issue. The rumen of a cow – the largest of its four stomachs, where a lot of the fermenting takes place – finds it has less to do, because the grain is already broken down and isn't as hard to digest as grass. The rumen, eventually, will atrophy and possibly fail.

When I feel the bite in the sun, and watch animals that barely have need to chew their cud, and think of the cattle in my care, particularly the dairy cows, I wonder whether they would be happy here. The answer is no, I don't think they would enjoy this environment much at all, though mine are mollycoddled with plenty of grass and close handling, and enjoy the temperate climes on our farm. When I ask myself if I'd be happy to feed my cattle genetically modified cotton residues, palm oil, coconut husks and the like – things well outside the normal diet of the ruminant in its natural state – it's a definite no.

But this system is designed with northern Australia in mind. Big cattle stations where the grass runs out each dry season can produce a lot of half-finished beef animals. Feedlots are the fix-it solution for the end of the animal's life. Even if it's tough, skinny, and not a great beef breed, feedlotting takes away enough movement, and adds enough fat, to make the meat palatable. It's meat for the bottom of the market, mostly. (Some does meet higher criteria, including Wagyu crossbreeds from this feedlot, and much of this high-end beef gets exported to countries such as Japan.) For the Australian market, mostly it's cheap, consistent, reliable beef that is available all year coming from this system.

Despite my reservations about this style of farming, the feedlots I've been to don't seem to have the huge health problems you get with cattle that are grain-fed and feedlotted their whole life – as they are in the United States, where much of the misinformation about grain-fed cattle comes from. In US cattle, sub-clinical rates of acidosis (in other words, cases where acidosis is present, but the cows aren't yet showing obvious symptoms of sickness) can be up to 37.8 per cent. And while Australian figures are hard to track down, up to 3 per cent of Australian dairy cattle get acidosis from the 6–8 kilograms of grain they are fed each day, too – so some of the health issues to do with their gut are not isolated to feedlots.

But I do wonder about the shade. I'm positively wilting even underneath the wide brim of my hat when I ask about it.

'The animals don't need it,' I'm told. 'You just get some customers who think it's a good idea. We have a big client who won't take cattle from here unless we provide some shade for the animals, so we do it for them. We only have it because of customer perceptions.'

'Oh,' I say, somewhat perplexed, as most of the cattle compete for space under the shade cloth, while I feel the full force of the sun on my back and head. 'They don't need shade?'

'No,' he replies, then hesitates. 'Sure we used to get a few deaths, especially in February [high summer] ...' his voice trails off.

Apparently the limited shade in the feedlot is only there to keep his large customer happy – not his cattle.

'Customer perceptions' – i.e. consumer expectations – are a powerful force. The feedlot owner, despite his reservations, had put shade cloth over at least some of his pens because of what his customers (possibly a Japanese importer, or an Australian supermarket) expected. Sow stalls (cage-like structures that confine a pregnant mother pig in a bathtub-sized space for its 115 days of gestation) are now gone from most of the farms that produce pork in Australia, thanks in part to the pressure exerted by supermarkets. Coles, knowing their customers – the Australian public – have higher expectations, went to the pig farmers and told them to remove the sow stalls or lose Coles as a buyer. It helps to be the customer who is one of Australia's two largest grocery chains.

———·———

Intensive animal industries have been getting a bad name. My aim, over the last ten years or so, in between my real jobs, has been to try to find out whether that bad name is deserved or not. After all, if you want to feed the world meat, you need to rear a lot of animals, using minimal feed, in the shortest possible time on the smallest possible amount of land. The arguments in favour of intensive animal production (or Concentrated Animal Feeding Operations, CAFOs) include some strong environmental arguments on the face of it. You can grow a bigger chicken using modern genetics, with a chicken that doesn't move around, in a smaller space, in less time, and hence using less feed, in a shed than you can outdoors. Since every bit of grain you grow to feed to animals has an environmental cost, efficiency counts.

But what I've seen – and I was shown the absolute best of the industrial animal system – doesn't always seem a good advertisement for meat. What is happening on those farms is unfortunately sullying the reputation of all farmers, and all meat eaters. What is done in our name, by people whose understanding of the complexity of animal welfare has, at times, seemed lacking, poor, or hardly in line with community expectations, is a sad indictment on an industry that sometimes values secrecy over social licence.

Consumer expectations are a wonderful thing. It wasn't chicken farmers who came up with higher welfare standards for meat birds in Australia. It was the RSPCA. They looked at sheds, at serious welfare issues, and came up with a strategy that could be applied to existing sheds to ensure capital costs couldn't be considered a barrier to existing farms. And when no chicken farmers put their hand up to improve welfare all on their own, it was pressure from supermarkets that got them on board. Supermarkets don't, by definition, care about animal welfare. But they know their *customers* do. So the RSPCA took the idea of higher-welfare chooks to the supermarkets, and got them to pressure the farms. Nowadays, about 70 per cent of Australian meat chickens meet RSPCA minimal guidelines for welfare. It's a great example of public demand driving change. The same can be said for supermarkets demanding the abolition of sow stalls, and perhaps asking for shade on feedlots in our harshest climes.

If farmers want social licence, if they don't want to be demonised, they will do things that their end customers want. It's a short-term solution to simply lock the farm gate – but as we'll see shortly, that's exactly what the most intensive industries are wont to do.

I delve more into cattle farming in Chapter 12.

CHAPTER 3

Killing the Mothers:
The Australian
Pig Industry's Dark Secret

'You've got pigs,' muses the feedlot owner as I am leaving. The camera is packed and the sound-recording gear is stowed after filming for the SBS television documentary. A long drive to meet a flight home awaits, and I've been given the hurry-up to get on the road twice already. 'They're great, so much fun, aren't they?' the feedlot owner adds.

I pause. This conversation suddenly seems odd, because on most things regarding animals this man – a fellow who lives and breathes livestock – isn't really on the same wavelength as me. We were walking around in 38°C heat looking at cattle in his care at his feedlot in central Queensland. Heat radiated off the laneways between the pens, where pale Brahmin and darker Angus crossbreeds stood staring at us, lazily swishing their tails at the flies.

Most of the cattle were huddled under the shade cloth that lined the centre of their pens.

So I was surprised when, at the end of two days on his property, he told me how entranced he was by pigs. Yeah, sure, I agreed. I adore pigs, too, for their curiosity, their sociability, their intelligence. We've become pig breeders and pig farmers because we reckon pigs are not only good to eat, but because they're a great animal to have around the farm. I turned back, curious to learn what it was he liked so much about the animals after which our farm is named.

I'm not a farmer by birth, or by training. I'm a chef by trade, drawn to the land by a love of ingredients, a love of flavour. For the last decade I have grown food for myself and family, on our piece of earth, Fat Pig Farm, in southern Tasmania, raising livestock, growing plants, trying to piece together a living from farming, and from cooking, and all the time conflicted about our relationships with livestock. It wasn't long after I got sheep that I first heard the farmer's cliché, 'If you have livestock, you have dead stock' – a less than poetic snapshot of what it means to rear animals. From the outside, this kind of talk can seem heartless, or cruel. To the farmer it's about being pragmatic. There's a lot about growing stuff that I'm keen to learn, and some lessons that have been profoundly tougher than others, mostly involving death.

Pig farmers are a curious breed, as curious as the pigs they rear. Many's the time I've bonded with people who have pictures of their favourite sow on their phone, as easily found as pictures of their kids. So I was keen to talk pigs after a few days of talking about the wants and needs of cattle. I thought I knew a lot about modern pig farming before then, what with sow stalls, and concrete sheds, and some footage of awful-looking conditions at Australian farms illegally shot and posted on the internet.

My newfound pig enthusiast buddy told me how he'd once set up a piggery in Queensland for an international company. Before they started the build, he visited one of the company's piggeries in Japan. It was state of the art, and biosecure. So biosecure that as you entered each section of the piggery – the breeders, the weaners, the birthing area, and so on – you had to shower and change clothes on the way in, and again on the way out. So, eight separate sections to visit meant that during a tour of the whole piggery, he had to have 16 showers in a day, because the piggery's owners don't want anyone bringing in a disease, or spreading any disease within the building. Obviously a disease that the pigs didn't have any immunity to, in such concentrated housing, could prove disastrous. So shower in, shower out, 16 showers a day.

I was interested, but faintly disturbed. It seemed a strange anomaly that we could have invented a system where animals are either so compromised in their health and immunity, or are so densely packed, as to require such high levels of isolation. Imagine a human city so lacking in basic immunity, or so poorly designed. But this was an ultra-modern, brand-new piggery. Technology at the cutting edge.

Once the Queensland piggery was built, they had to stock it with their first pigs. Commercial piggeries use farrowing crates, so at this point in the story I pictured healthy sows farrowing (giving birth) within this intensive system – albeit in farrowing crates (which I don't agree with, and which I'll talk about in Chapter 14) – suckling large, healthy litters of piglets.

'We wanted the latest in pig welfare,' my pig buddy is telling me. 'And you wouldn't believe what we did,' he added. 'On the day the sows were due to give birth, we got all of them, and brought them up to a building just outside the piggery.'

Then, in the same excited tone, with the same enthusiasm, he told how they killed the lot of them, cut out the piglets, plonked them into sterile wheelbarrows, and bundled the newborns into pens inside the new piggery.

By this stage of my personal journey, I'd heard some pretty grim tales. I'd seen and been responsible for more death than I ever considered possible. But I hadn't heard anything about this. Was this a one-off? Was this bloke just making stuff up to try to shock me? Is this killing of the mother pigs common practice?

I almost thought I'd misheard this as a valid way of stocking. So I looked it up. These, in industry speak, are called 'minimal disease pigs' or 'specific pathogen-free' (SPF) pigs. You used to be able to find minimal-disease pigs, and a description of the euthanasia of the mother pigs post-hysterectomy, on the Queensland Department of Agriculture and Fisheries website – as though it was just another thing that humans do, like provide water to calves, or a pat to a dog. Interestingly, during the editing of this book, that same website removed its reference to minimal-disease pigs, and now references 'hysterectomy' in a more obscure way, but still to produce low-disease piglets. While technically accurate, hysterectomy of a sow to produce SPF pigs means the mother pig will be killed immediately prior to the operation to remove her uterus, with the piglets then moved away from the mother's body to avoid disease transmission. I think it's telling that the specifics of the euthanasia have been removed from the website, making it harder for the general public to see what probably is, in most people's minds, a questionable if not downright condemnable practice.

If you've ever had a baby or been present at a birth, or even been around a breastfeeding mother, you might have heard of colostrum, the vital first early milk. The colostrum of pigs contains 25 per cent protein, mostly in the form of antibodies for immunity. It's vital that piglets obtain some in the first few hours – and if they don't receive any in the first 24 hours, their immune systems are never quite the same. Dairy farmers know this. Midwives know this. Most people who deal with animals know this, obsessing about how vital colostrum is for providing immunity. (In an existing piggery, the piglets may be given an adoptive mother for suckling if an SPF mother, who has just given birth, exists in the facility.)

Minimal-disease pigs? Those not connected to the high-pressure, high-turnover, industrial nature of the pork industry might call them immuno-compromised pigs, lacking that all-important transfer of immunity from a normal vaginal birth. Piglets that are so weakened immunologically that they need to be placed straight into what is essentially a humidicrib – before being tended in a 'laundry basket', as the government website used to describe it, ready to pop into the piggery. Can we really have invented a system where missing out on colostrum is part of the management of the animal – and who could possibly think it would be a good thing for pig health? Let alone brag about it.

Ultimately, no commercial piggery grows pigs only for its own consumption, so it does make you wonder about the ethical mindset of someone who thinks nothing of killing tens of mothers on the eve of giving birth, and whether this kind of farming system could ever pass the pub test.

This is farming for the farmer, not the animal. And even then it doesn't seem to make sense. Everything we know about immunity counts for nought in this system. Pigs evolved to drink their mother's milk with all its antibodies, to root around outside and get exposed to bugs and build up their resistance. Our industrial farming system has created these fragile, weakened animals through the production system, through our own vested interests. We incarcerate pigs in such confinement that they are highly susceptible to disease, much as humans encountered in badly managed urban centres when we first congregated together. But instead of thinking of the big picture, such as the flaws in the system and how a pig could be so lacking immunity, and how to fix that problem at the outset, we deal with the end result. We try to fix the symptom, not the true cause.

The causes of disease can be many. We know, from the human experience and studies on animals, that the more we mingle with larger and larger groups of our species, the more disease we're exposed to. If we aren't very careful with certain things – effluent, garbage, the food source, etc. – this disease potential expands enormously. And we also know that the best way to build resilience is through exposure to tiny amounts of a huge variety of bacteria from the normal vaginal birth onwards, to drink colostrum, and be subjected to the elements, dirt and the like.

We know the immune system can be compromised by the use of antibiotics, not just from its use within this generation, but also within previous generations. The more antibiotics your mother and grandmother were exposed to, the more likely you are to have issues with your own immune system. We've co-evolved to exist with bacteria – over ten thousand species of them live in and on us, comprising about 2–3 kilograms of our adult bodyweight – and without them we'd die. Bacteria can be our friends in the fight against disease, and it's when only a handful of species get in the wrong place, or in the wrong numbers, that potentially makes us ill. Besides the bugs on and within us, we also need to have a diet with a wide variety of food sources to ensure the immune system we build is as resilient as possible. A pig's anatomy and biology, which is remarkably similar to a human's, means a lot of this is also relevant to them.

Now, not allowing a pig a normal birth and to suckle; by using antibiotics as a prophylactic; by using a limited grain-based diet;

and not letting the animals out on dirt, instead confining them in very close quarters – these are all good ways of creating a problem. That's the *cause*. The *symptom* is disease that can rip through a large shed in a short time. The *solution* humans have found is more antibiotic use, more biosecurity, more concrete, less use of normal food waste and more use of grain, and more confinement.

All solutions to a problem we have caused, and none of them in the individual animal's best interests in the longer term.

———·———

I do adore pigs. We have a handful on our farm in southern Tasmania, pigs that live outside and provide for our table and for the table of those who visit us. Our pigs live and die at our hand, and we're no strangers to eating meat. But hopefully our pigs also get some joy in their lives, as they dig, and wallow, and express their pig-ness – the instincts they have to nest, snuffle and socialise.

I know that they damage soil, and can be difficult to manage in our system, and I'm also not convinced free-range pork can be done sustainably and feed the demand for pork.

When we filmed *For the Love of Meat* for SBS, we approached several large, intensive piggeries, including Australia's largest, Rivalea. We hoped to enter the farms legally and see an intensive system for ourselves, and through the medium of television, show that pig farm to consumers of pork. Having watched a lot of animal rights videos about the conditions on intensive farms, I was prepared for the conditions to be grim. But I was also fairly sure that the illegally shot videos I had seen were of the worst possible cases, the rotten apples of the pig industry. Sure, a farm with 45,000 sows wasn't going to be anything like the storybook conditions on our little farm and some free-range farms, but surely it wasn't going to be that bad?

Surely, the large-scale farming done in our name was going to have some ethical frameworks that consumers could be comfortable with?

Surely, it couldn't be all that bad?

Stories like the euthanised sows and minimal-disease pigs, however, make me think it probably is as bad as you can imagine. And without oversight, without accountability, without opening their doors to the people in whose name such outrageous acts are done, the pork industry does itself a disservice.

I'd like to think farms aren't killing animals unnecessarily, aren't trying to retrofit flawed systems by treating their livestock not as living creatures, but rather like mechanical moving cogs in need of a spare part or a bit of a grease and oil. But you wouldn't know, because you're not allowed to know. Farms don't have to give us, their customers, access, or even information. They don't have to show us or tell us what goes on behind their locked gates and large biosecurity signs. The intensive animal industry operates on the three main principles of secrecy, secrecy and secrecy.

In a world where two out of every three domesticated animals reared for the table is raised in an intensive system (commonly called factory farming), surely we need a moral framework. As meat eaters, surely it's incumbent on us to know how animals have lived and died solely for our benefit. Some may choose to turn away. Humans are very good at rationalising what is in our interest, and some will simply not want to know. And that's fine. But some of us are conflicted, and do listen to our inner voice, even while still choosing to eat meat quite consciously.

We're conflicted as individuals, as a society, as a species, with the things done by and for us to the land, the environment, to the animals in our care. Surely, then, we need to have guiding principles.

I've met pig farmers who say they're so proud of what they do, yet they won't open the gate to confirmed meat eaters because consumers may misunderstand what they see. Hamstrung by the need to produce pork at unprecedented low prices, these farmers worry that what happens behind the shed doors would frighten their customers.

I wonder, though, if these people in the industrial meat industry, who probably meet the definition of 'farmer' far more than me, have every facet of production in mind – including the animal's needs and wants, the environment, and consumers? They seem to have a profoundly different world view, and hence animal view, than I do. I wonder if they have the same standards as society, even if these differ from mine? Does what they do have social licence, despite my reservations? Can these people hold their heads up high amongst the meat eaters of the world and proudly proclaim that their farming credentials and systems meet society's expectations? Will they actively tell you about minimal-disease pigs, and feel proud of what they do, in your name? I suspect there's a reason that this kind of animal husbandry isn't common knowledge.

The sad fact is, if you eat conventionally grown pork in Australia, you are probably complicit in treatment of animals that most of us wouldn't condone. Like it or not, all of this is done on our behalf – and the situation is probably even worse with imported pork.

And like it or not, as meat eaters, we, more than those who choose not to eat meat, should have some say in what we think is ethical and not.

Sadly, the pork industry in Australia likes to keep its secrets, and unless we make our displeasure clear, they will continue to do things that we, as a community, would never knowingly countenance. The only way it will change is if we demand more of our farmers, more accountability, more openness, more social licence. The only way the pork industry will change is if we force it to.

CHAPTER 4

160,000 Deaths:
A Medium-Sized Operation

'You don't want the chicken?' Both the chicken-plant owner and his publicity person are looking at me with disbelief.

They're clutching two bagged, plucked and drawn chooks, freshly killed within metres of where we stand. These two nude-blonde birds are just two of about 160,000 chickens that have met their maker today. Met their maker almost literally, because chicken farming is vertically integrated, meaning that the company who farms them (or has the growing contract on a farm) is almost always the company that owns the abattoir.

And having watched many of the 160,000 birds that died that day hanging from the gallows, no, I don't want chicken for dinner. While the smell of the growing shed will find a special part of my brain to hide in for goodness knows how long, the smell of the killing floor will, I now know, live with me forever.

I'm no stranger to chicken death. I've had to put down our own chickens for ethical reasons when they're ill or old. I've cut the heads off chickens so that the birds can grace our dinner table, a practice I now know – with shame – does not kill them immediately. I've fought to overcome the smell of the deceased birds on my hands and clothes after a day at the plucking table. But this factory was a place that haunted my dreams in the days following the visit. Years later, it haunts me still.

And these were the good guys. Proud of what they do. They were friendly, helpful, and above all, were happy to show not only me, but also the rest of Australia, how they rear and dispatch chooks. But still, a lot of chickens, in anyone's language, were due to die that day. Apparently this is a 'medium sized operation' in Australian terms – a tiny operation compared to some in the Americas and China. Some 11–12,000 birds processed per hour at this plant doesn't seem tiny to me, and yet this is a fraction of what really big abattoirs can handle.

One of the four conveyor belts – one that held dead birds upside down for much of the processing and chilling – was 4.2 kilometres long. Bird after bird, alive at the start, is whipped through the process. Gassed, bled, plucked, gutted, chilled. The whole scenario is a graphic example of human ingenuity.

The birds, snatched from their sheds in the dark of night, are held in crates in a loading bay with eerie blue light which, apparently, calms them. When it's their turn to die, each crate load is moved onto a conveyor belt that passes through a gas chamber, where high carbon dioxide levels leave the birds unconscious – 'stunned' is the term – ready for their throats to be slit. You can't see inside the gas chamber, but I'm assured the birds just go to sleep.

As the birds emerge, motionless, they are strung up by one hock on a clip on a conveyor belt, and swing off past the rotating knives. I watched most of this from a glassed-in room at first-floor level.

The birds were bled from the cuts to the neck, checked to ensure they were bleeding out properly by a human wielding a razor-sharp blade in full HAZMAT suit and hair net, ready to do it themselves if the automated knives didn't work. The conveyor belt snaked away, around the edges of the room. Chickens about a foot or so apart, all dangling from the leg. All white, bird after bird after bird.

The smell of the factory lingers outside in the carpark, even before you enter. It's stronger still in this glassed-off room. Dressed in full covering gear, I went down to the kill floor, where you are exposed to even stronger smells – of wet feathers from the plucking, the metallic tang of blood. The distinctive odour of chicken guts. It's noisy, it's oppressively humid. I hear the rhythmic *shing, shing, shing* of the fine blades cutting off the feet. It's warm in these rooms, where the bodies of all these birds pass by overhead then disappear into rooms beyond.

At one point, on a gangplank, with now-nude chickens making their inexorable march along the conveyor to the chiller and back again, I feel giddy and sick. Inside the chiller, the chickens would pass one way, then another, higher and higher, doubling back on themselves as the conveyor belt whirred on and on and on.

Standing here, underneath so much death, I start to wonder at our hunger for meat. Our hunger for this meat in particular, a small animal. For so many dead birds to make so many cheap meals to feed so many people who really aren't lacking in high-quality protein. I start to feel the pressure our voracious appetite for meat is putting on the world, on the birds, on the farmers, on the workers on the kill floor.

This is a well-managed, state-of-the-art abattoir. It's a mostly automated process where the humans involved are often just checking for signs of life, ensuring the throat has been efficiently slit by the rotating blades, looking for any indications of a bird that was stressed before death (pink, dehydrated, or similar) and is hence discarded. Humans are there double-checking the evisceration, where a machine sucks the guts from the birds.

Without knowing why, I feel pity for those workers, whose days are spent having to do things many of us couldn't do ourselves. Work that we feel might demean the rest of us. That demeans even us meat eaters, in whose name they do these dirty deeds.

What happens in this kind of factory – and it *is* a factory – and all those other factories that *weren't* open to my visits? All that work done on such a scale to feed a chicken-hungry nation? Watching the scale of all this death, at a *medium-sized operation*? I'm starting to think it demeans us all.

———·———

I pity the chicken. An animal plucked from relative obscurity to become the fastest-growing worldwide meat of choice. I pity the chicken for its docile nature, its ability to grow fast. Its innocuous, often bland flavour. All these traits have ensured the chicken is the most intensively farmed animal on the planet.

According to the peak body, the Australian Chicken Meat Federation, about 650 million chickens are consumed in Australia every year, the vast majority produced by just seven companies.

Of these 650 million birds, all but a modest few thousand are completely white in feather, and virtually all the genetics come from two monolithic breeding firms based overseas. They're all Cobb or Ross genetics, named by the companies that own the breeds, designed by years of breeding and millions of dollars of research to get from the egg to the pot in about 35 days, to be bland and tender, and to have white feathers, so you don't see any dark pin-feathers. (Dark pin-feathers are considered unsightly, the same way dark hair on pork is considered unsightly on the outside of the meat.) These white-feathered animals are designed for the farming system that we've gradually perfected to keep them in – sheds crammed with 16 or more docile-natured birds per square metre, with an ability to convert high-protein feed into chicken flesh at an astonishing rate.

Chickens, originally, are a jungle fowl. Probably from India and South-east Asia, it is thought they were first kept as an object of beauty and sacrifice. Chickens also have the misfortune to produce an egg that is palatable, and meat that we enjoy. That, along with the characteristics that allow them to be manipulated as livestock (friendly, able to be handled, not frightened of humans), has driven chicken to be the number-one meat in Australia – way ahead of beef, a long way above pork, and streets ahead of what many of us think of as our national meat, lamb.

We've domesticated chooks for about 8000 years – far later than dogs, and much earlier than cats. For a long time, chickens, like pigs, were kept around farming communities because they not only provided for us in terms of meat and eggs, they also recycled food scraps and other waste that humans couldn't eat. They are one of those miracle animals that take stuff that we can't digest, and turn it into high-quality protein. Their small size meant that they were traded often and quickly spread around the globe, including every time ships made voyages further afield.

For a long time chicken meat was a luxury. Kept for eggs, and the occasional roast dinner, the farmhouse chook was of more use alive than dead. Only the roosters met an early fate, because they crow and fight a lot, and you only need one rooster to get fertile eggs. But the hens were kept as long as possible, as they still are in developing nations like East Timor, where the eggs from the hens provide way more meals than the life of the bird taken young. A good 250 eggs in a laying year from a decent chook, or one 2 kilogram roast bird?

Your pick. Even if you did want the meat, it might take four or five months to get a bird of size – so roast chook was a luxury reserved for Christmas and the occasional Sunday.

Fast forward to today, with modern breeding, temperature-controlled sheds, and the advent of antibiotics and fortified feed that negates the need for the birds to create vitamin D from sunlight, and we have an industry of mammoth scale on our hands. Tell me the cheap takeaway or salad joint you've been to lately that didn't have at least one chicken option on the menu, if not half the options?

———·———

One of the stated aims of the chicken industry is to produce a bland, tender meat. What they want is meat that won't offend anyone (flavourless, or at least insipid), and isn't chewy. Tenderness isn't usually a problem in a bird that is well below full feather age and dead after four or five weeks – one that can exist in an environment where the lights are never turned on enough for the birds to be fully alert, but where the lights are, for 23 hours a day, just bright enough for them to find feed. This leads to lazy, dopy birds, whose motion is slowed, whose feed conversion rate is very high (a plus in environmental terms), and whose muscles are under-utilised, so their flesh remains bland and tender even as they age.

A *Sydney Morning Herald* tasting panel in 2010 declared that chicken should be 'moist and tender with a clean flavour that doesn't really taste of anything'. The head of Ingham's declared at the time that 60 per cent of the meat they sell is now marinated to make it more tender, adding salts and phosphates (to retain more moisture) and 'flavour profiles'. As a cook, I can tell you that if you have to marinate an already tender meat, it probably means the meat needs something else to make it taste.

As a farmer, I can tell you that good meat, carefully reared, can taste great all on its own.

————·————

The chicken and the rooster are strong cultural icons. We put them on crests, laud them on flags, use them as symbols of fertility and in branding, and yet debase them in reality. It's the cheap meat of choice – ubiquitous at takeaways, at deli counters, in sandwiches and salads. And it's there, in large part, because it's now very cheap.

It's cheap because we grow them so fast – about three times faster than we did 100 years ago. We grow them intensively, with up to 42 kilograms of live birds per square metre of floor space, and with quite a bit of chemistry going on in the feed (see also Chapter 13, on antibiotics). We use temperature and humidity and air quality controlled sheds to make chickens grow so fast, too – 650 million times a year.

According to a research paper in the journal *Nature* in early 2018, chicken growth is a mastery of manipulation. The modern meat chicken, called a broiler, 'consumes 3.2 kg of feed over 35 days to achieve 1.8 kg of body weight, compared with more than 20 kg of feed over 112 days to attain the same weight in the 1920s'. In part, this can be attributed to the growth-promoting ability of antibiotics, but even ignoring that fact, this means a modern chicken uses about 15 per cent of the feed and takes less than a third of the time to reach the same weight. We can get a chicken to size in as few as 30 days today, down from 84 days in the 1960s.

At one stage an Israeli company was trying to breed a featherless chicken – because in our modern, fast-growing, indoor system, feathers are just an annoyance. Kept in finely controlled conditions where the temperature is constant, and the humidity well defined, feathers are an expense at processing, and a possible contaminant. The featherless chicken idea has gone quiet lately, possibly because images of it would offend many, and the look alone may not pass public examination.

The featherless chicken is just one of the many ideas people are thinking of. There's research into whether we can grow chickens that feel no pain, or experience no stress. We are still on the lookout for ways to debase the original chook even more than we have already.

This constant intensification comes at a cost. These birds grow so fast that their short legs are weakened. Their hearts can't take much stress. A modern meat chicken is designed for modern supplemented feed, for modern sheds, for a short, dull, sedentary life. When I was researching my documentary, I asked the chicken producers and researchers if we'd already pushed the bird as far as it could go – perhaps even too far in terms of animal welfare outcomes. Their response was an adamant 'no'. We can push the breeding even further. We can make chickens grow faster, cheaper. We can produce even more animal tofu because we're clever, and people want cheap, bland, tender 'meat'.

So what do we sacrifice with this fast-growing, lower greenhouse-gas emitting chook? Well, one of them is flavour. And nutrition. The fat profile of the modern hen is weighted more to omega-6 fats than the more beneficial omega-3s you get from an animal given leafy greens rather than pure grain. The modern chicken has way more kilojoules per gram, and a lower protein content than its pre-industrial farming counterpart. The lazy, inactive bird on a grain diet is fattier, lower in protein, and relatively tasteless.

In Australia we farm chickens this way because we can. We grow white, bland, fast-fattening chooks because we can. And because so many people buy them.

There are other breeds of meat chickens out on the fringes. Sommerlads are perhaps the most high profile of the alternative breeds, developed by a couple of the same name who were appalled by the industrial system and moved away from it. They changed tack because they wanted to produce commercially viable birds that are hardy and are happy to free range, but also have way more flavour. I talk more about pastured birds in Chapter 24, What Is Good Meat?, later in the book.

Not every farm with white birds is pushing them to grow super fast. One farm I went to, Inglewood Farm's organic chicken farm and 'processor' (read abattoir and cutting shop) in Queensland, like all organic producers, has to grow chickens without antibiotics and fatten them slower. This farm got into the game of growing chickens because they already grew the feed. Organic grain can be up to five times as expensive as conventional grain, and some of the additives, such as oregano powder (which is a natural anti-microbial, negating the use for routine antibiotic use), can cost up to $40,000 a tonne, or $40 a kilo. The hens are kept at a maximum of 25 kilograms of birds per square metre (about 60 per cent of the stocking density of the most intensive sheds), plus they get access to the outdoors. The final chook, despite being the same breeds they fatten in every intensive, conventional shed, is slower growing, and at least twice as expensive, sometimes even more. You do, however, get what you pay for, in terms of better texture, better flavour, a happier bird, and better environmental outcomes from the growing of the grain.

———·———

The advent of cheap chicken has spawned other industries. Mushrooms are often grown in the rich waste from chicken sheds.

Market gardens may get their compost enriched from the poultry waste from so many birds. Salmon farms in the local waterways near our plot of land are made more viable thanks to the cheap protein, feather meal and fat that is sourced from intensively reared poultry. (That's right, salmon are fed chicken. They're also fed blood and bone, aka meat meal, from mammals, too.) In the United States, feedlotted cattle are routinely fed poultry litter (the stuff from the bottom of a chicken shed: a mix of chicken poo, undigested food, feathers, hay and the like) – even though cattle, in nature, are not really known for their consumption of feathers.

It's interesting to look at why the United States thinks this is a good thing. They've obviously not had the mad cow scare Britain experienced, and the dietary needs of cattle are considered by nutritionists charged with finding high-protein feed at the cheapest cost – hence the use of poultry litter as feed, because it's cheaper than other sources of protein. But if you've ever been in a chicken shed packed with 75,000 birds, you'd be quite surprised to think a ruminant (essentially a vegetarian like a cow), would be fed something containing excrement and feathers – something that smells toxic and is banned in most countries in the world. It's telling to look at what one educational facility, the University of Missouri, has to say about the practice:

> *Poultry litter should not be fed to dairy cattle or beef cattle less than 21 days before slaughter. The reason for this prohibition is that the residues of certain pharmaceuticals used in poultry production may be present in poultry litter.*

And if you are surprised that American cattle growers feed their cows poultry litter, despite the poultry litter possibly containing residues from pharmaceuticals (including antibiotics), that's because you're not in the intensive animal industry. It's only the rest of us that find it abhorrent. The University of Missouri notes the issue can create a stigma for the beef industry, and that the growth of pathogenic bacteria and accumulation of pharmaceuticals in the food chain must be managed. I'd have thought avoiding these in the feeding stage might be a good start.

———·———

It may seem sacrilege to those who have fallen in love with the lack of flavour in modern chicken meat, but surely soft, insipid meat, often sold marinated, and as a great vehicle for other flavours – surely that is merely animal tofu? Do we really think 650 million lives should be created and lost each year so we can have something that tastes like the soy sauce and lemongrass we soak it in? Can we justify the sheds reeking of ammonia to rear chickens, and justify the jobs of those who slaughter them at our command, when all we want is chicken that is resonant of the tomato and oregano we put on it, or the crumbs we coat it in?

The answer, obviously, is yes. We, as a nation, *do* want chicken without flavour. We *do* want something with a tiny bit of chew – but not too much chew. Otherwise we and other nations would have rejected the modern chook as a pale imitation of the real thing, and would simply find that the mass production of them insults our sense of self.

———·———

The Huon Valley in Tasmania, where my family is based, is full of backyard chicken fanciers. There are plenty of people to bond with over your chook pen, your laying boxes, your Australorp, Barnevelder or Rhode Island Red. And there are lots of people who lop the heads off their own chooks to then feast on the meat at the dinner table. And this meat, I have to say, while sometimes on the more resilient end of the chewing spectrum, can be amazing in flavour. And guess what? A thing I've learned, since growing, rearing and dispatching our own birds, is that a good-flavoured chook is a thing of wonder. It's almost as rare as, well, hen's teeth. Or at least as rare as they were in our grandparents' day, when a chicken only died for feast days.

We find that a slow-grown, old-breed chook satisfies more, tastes more, and we eat less of it. One chicken can feed our family for four or five meals. More if we use it judiciously.

That flavoursome chicken, the stuff I dream of for all the months we don't have our own chooks to eat, isn't likely to come your way anytime soon. Probably because it costs a lot more to get a chicken like that – and because so much of the industry is still cloaked in secrecy.

I often think I'm too naïve, believing that people should grow animals in a way that suits their moral compass, their ethical standpoint. And some farmers *do* grow things based on standards they uphold of themselves and their staff. But not some companies, such as Baiada, Australia's biggest chicken producer, who we'll meet later in the book. If they did, they'd welcome scrutiny. They don't care whether an animal has the ability to express its instincts. If they did, they'd be happy to let a film crew inside the sheds. They don't care if a bird lives a short, unfulfilled, rudimentary life in a shed that stinks of ammonia. They don't care that the breeds they used for a while had a big problem with leg breakages; all that mattered to them was their balance sheet. These are the people who can watch even *more* than the 160,000 chickens I saw go to their deaths and still go home and eat chicken for dinner that same night. Perhaps, because of fast-growing modern chicken's lack of flavour, there's little association between the smell of the killing floor and what's on the plate.

That kind of farmer smells the shed and smells the processing plant and thinks it smells like money.

So what does the thinking meat eater do about chicken? After all, standards like the RSPCA's do increase the welfare of millions of chickens each year. Even the free-range farm that I talk about visiting later in the book, run by Baiada – a company that cashes in on the misery of millions of intensively indoor-reared chickens – even on *that* farm the free-range animals themselves do live a better life.

Some Buddhists suggest if you're going to eat an animal, make it a big one. Every time you eat a roast chook, an animal has lived a definitely short, possibly miserable, and almost certainly unsatisfying life in which it is barely allowed to express its natural instincts. For every two chickens you eat, another chicken probably didn't make it – so in Australia alone it's probably closer to 1 billion birds that are culled or die each year, from the time they hatch to the time they hit the table.

Every time you eat lamb or pork or beef, you're only eating part of an animal. One steer can provide about 70 roasts of beef or the equivalent. When you eat chicken, the taking of a life only provides a meal for about four people on average. A steer can make about 280 meat-rich meals.

I know, sadly, from experience that what I say and write may have little impact on the way Australians think of chicken, and probably a negligible impact on the number that sell. I know that what we believe in is often sidelined when we smell the dirt-cheap roast chicken at the deli, or when we buy the least expensive fresh bird in the meat section. But I can't sit idly by and not voice concerns about how those chickens are raised, what kind of life they lead, and the ramifications not only on the animals themselves, but those who have to work in such settings. I have to voice my concerns about the motives of the people who run these operations, their inability to see birds as worthy of some kind of meaningful life. I worry about that attitude's repercussions not just on the birds, the workers and the local environment, but also on the rest of the world as a whole. Chicken farming, the way we've pursued it, is a recipe for disaster. As we'll see in Chapter 13, about antibiotics, that disaster could well be unfolding as I write.

I wanted to talk to those who bridge that divide between the farm and the home kitchen. Those charged with the pivotal moment, when an animal ceases to be an animal, and meat comes into the world. I wanted to understand what kind of person kills for a living, and just how they see not only the animals that die at their hand, but also how they view the world that asks this of them.

I don't recommend a visit to a chicken-processing plant if you want to maintain an appetite for the bird. Not everybody should be on hand when an animal dies. But everybody should know just what it means, because death comes, and it comes to us all – plant, beast and human.

CHAPTER 5

The View from the Killing Floor

There's a moment in the production of meat that is more important than any other. A brief window for all that has gone before, and all that comes after, to go terribly wrong.

It's leading up to, and at the moment of death, that has far-ranging impacts on the quality of the meat we eat. A good life, and good butchery and cooking, can't undo a bad death.

Meat from an animal that has been stressed for hours, or stressed for seconds in the lead-up to the final moment, is lesser meat. No amount of good animal husbandry, deft butchery or kitchen wizardry can make up for the fact that the time of death is where things can go irretrievably awry.

It's probably good that things work out that way, because it means even those who aren't really into animal husbandry in the way others are – who don't think an animal needs to express instinctual behaviours – even those farmers are still interested in a low-stress final day, and a very quick and painless death.

This moment, though, when the animal dies, is shrouded in secrecy. Abattoirs are loath to show people the process around death, afraid it will cloud their judgment. I certainly don't think everybody needs to witness what happens at the end of an animal's life, but I do think an animal's death is something we need not be ashamed of. If we are ashamed, perhaps we need to think more about what we do, how we do it, and whether we can do better. As the saying goes, if abattoirs had glass walls, there'd probably be more vegetarians, and that probably wouldn't be a bad thing.

But at the end, there are people. Always people. Abattoirs are designed by us, to help workflow and health and hygiene, and more and more they're being designed with the animals in mind. And at the pointy end of all designs, despite many automated systems, there is always a human. What's it like for that person – the one charged with the responsibility of taking the life of multiple beasts so that we can eat meat? What does it feel like, doing the dirty work that most of us don't have the stomach to do for ourselves? How does it feel to work in an industry cloaked in secrecy, where parts of society don't just disagree with what you do, but actually disdain what you do, and actively try to disgrace your work and shut you down?

I thought it best to sit down with a slaughterman to see just what he thinks about doing a job that most of us wouldn't countenance.

Like most people in the industry, David Stephens didn't think he'd be a slaughterman when he grew up. With a face lined by years of hard work, and laughter, David has fists the size of saucepans, and arms as thick as tree boughs. He's slow to talk and not much quicker to judge, and I've known him, and have had work done by him, at his small-scale slaughterhouse and later as a mobile butcher, for a decade now.

At the end of his career, having sold his abattoir a few years back, he's not completely convinced that it's a good thing he chose the path he did. But he's pragmatic, as country folk usually are. Born to the owners of a dairy farm in New Zealand, the electricity only went on in his lifetime. In those early days the milking machines were run off the tractor, his parents sold the cream and fed the skim milk to pigs. The farm ran up to 180 cows at any one time over the years, and 24 sows. David has, for most of his life, been around animals.

Chance, and love, saw David move to Tasmania, where his wife Rita and her family ran boarding houses. With an escalating meat bill to feed the boarders, David and Rita decided to rear their own pigs, and pretty soon ended up fattening up to 600 pigs at a time in sheds, behind the village of Franklin, just across the Huon River from our farm. David wasn't overjoyed with the way his pigs were being handled, however, in that all-important time post-farm and pre-slaughter, so he got into the abattoir business himself. At the time, Tassie fielded about 80 slaughterhouses, with many small operations all across the regions. With new regulations that came into effect, that number quickly shrank to only about 30 in the state. There's now only a handful of abattoirs left, and most animals must be trucked a long way before slaughter.

'I never really killed anything until I had an abattoir,' admits David, despite his rural upbringing. It wasn't until he actually owned the slaughterhouse that he had to do the deed himself.

'I was working in our own place, killing about two sheep an hour, and then these slaughtermen came down from up north and did 50 in an hour. It sure was a learning curve.'

As the abattoir owner, David was also the one on the kill floor. 'That's all I did,' he says. 'Over six or seven years, I reckon I did about 50,000 cattle … and I forget how many pigs.' He does some quick calculations on his fingers: 'Say 100 pigs a week, up to

400 around Christmas,' and soon it's pretty clear that about 35,000 pigs died at David's hand. 'It's a bloody lot of pigs,' he muses, without relish.

This amount of killing, does it ever get to him?

'I think it's ...' his voice falls away. 'It *does* get to you. Rita and I, we don't eat beef at all, really, now; it changed the way we eat.'

David's aversion to beef started with his experience of one particular animal, a Lowline Angus that had a certain look in its eye. 'That's when I started to struggle, after that,' he says. 'It was the most beautiful-looking calf, and for some reason, the way it looked at you ...'

Interestingly, David can vividly remember that time that one animal, just one of the 50,000 or so that he sent to the grave, looked at him, and knows the impact it's had on his life. Or maybe the effect of that calf was the accumulation of all the impacts of the job.

One of David's sons, who worked the kill floor for much of the time, also doesn't eat much meat at all these days.

'The killing really got to him,' David says, and perhaps the regret of a father looking back at his time on earth is what I hear tingeing his voice. His other son, who still does some home kills as a mobile butcher, can no longer bear to do the slaughter alone. Todd may wield the gun and stun the animal, but David is still on hand to bleed the animals out. 'Todd shoots it, and then he just walks away, and then he can handle it better,' David explains, a father's concern for his son's emotional wellbeing obvious beneath the tough exterior.

Is there a strategy to cope with the demands society places on a slaughterman?

'When you walk onto the slaughter floor, your mindset seems to change. You just do your job and that's it. It's really hard to explain. I'm sorry, I can't help ya.'

David has strategies for killing. For dealing with the grim task that society – and his own life choices – has asked of him. He focuses his mind. He sheds as much emotional attachment as he can. He exists in the moment.

Has all this killing changed you, I ask?

'Christ yeah, course it has. Well, I just, I appreciate life more, after killing all that stuff,' he says. 'I'll go fishing with Rita's brother, but I'd rather not go, and I used to love fishing,' says David, wistfully. 'When I go fishing, I now have to use the same mindset as when I was on the slaughter floor.'

David seems soft hearted for a man who has seen so much death. Maybe it's *because* he's seen so much death.

'Often you'll see these fellas going shooting, and the first thing that comes in my head is "Stupid bloody rednecks" – and yet that was me, once,' David recounts. 'I'll tell you this: seeing what I've seen, you appreciate life a damn sight more.'

Are there any regrets, leaving farming and running an abattoir?

'I wish I hadn't done it, now,' David says with some sadness creeping into his voice. 'I know it sounds funny to say I wish I hadn't done it – but I don't regret it, if that makes any sense. I've done it, and the only thing that made it for me was going mobile [for a few years David ran the mobile butchery business his son Todd now operates], and meeting all those beautiful people who I did all this work for and up to then had never met.'

Is there a stigma around killing for a living?

'Definitely, yeah,' David reckons – especially when he's in a crowd.

'You'd be careful who you'd tell. Some people don't talk to you, and you know why they're not talking to you. It's odd that some people don't like you because of what you do.' He seems puzzled by the fact he can be made to feel uncomfortable by the very people who ask him to do the act of killing on their behalf, the people who choose to eat meat. 'That's just part of it,' he admits, without judgment.

When I ask David if there's one thing the general public could understand better about what happens in an abattoir, he doesn't hesitate.

'Oh, look, 99.9 per cent of it is just so humane. You get a few halfwits who work there, like you get in any industry, and they're the ones that seem to get publicised, aren't they? And all the good people aren't talked about. The animals are really looked after well. They really are.'

Like so many of the farmers I've met, so many of those who've thought about the meat they eat, David has a line that shouldn't be crossed – his own beliefs around what is good, what is right, and what community standards to apply to livestock in our care.

'There are chook farms that, really, are so bad they shouldn't be around. Like those rednecks I was talking about. Some things just shouldn't happen.'

David finishes our conversation with the same kind of sentiments I've heard from those who choose not to eat meat, any meat. The kind of talk that seems surprising to hear from someone who, for more than a decade, was paid to kill animals so the omnivores in the world don't have to. A telling statement from someone who knows the end of an animal's life as well as anyone.

'Animals have got feelings, for god's sake.'

CHAPTER 6

Where are the Animal Lovers?

While we all have our own moral compass, our own ethical viewpoint with its ambiguities and biases, it's never a bad thing to be challenged on what you believe is right.

I remember walking around the phenomenal and scary Qingping wet market in Guangzhou in southern China in late 2000. A local was showing me the plethora of available produce to take home to eat, most of which were still alive. There were scorpions in buckets, frogs in tubs, civets behind bars. I counted 16 trucks unloading turtles. There were tanks of snakes of all sizes, skinned dogs hanging up in doorways, and a man wrestling a writhing hessian sack under water. Drowning a cat, my guide explained. In cages were innumerable chooks and ducks and pangolins and the like. The eating of any creature whose back faces the sun – meaning anything that didn't walk upright like a human – was very, very strong in those days, and may be still. In the year 2000, over a four-day period, local authorities raided markets, feedlots, hotels and restaurants in Guangdong province and rescued 17,582 wild animals that were protected under state laws. From the tiniest of birds, to winged insects, from reptiles to mammals, there's little that wasn't for sale at the Qingping market.

I won't say I was unmoved. It was all a bit confronting really. I had once toyed with vegetarianism as a student, and while it'd been a few years since I'd started eating meat again, the graphic deaths of so many different animals in the market was shocking. The smells alone were far from pleasant, the whole thing a ghoulish scene that could give a city boy from Australia nightmares.

Having been astonished at the array of living things that could hit the dinner table, I asked my host what his favourite meat was.

'Cat,' he said without hesitation.

'Cat?' I responded, a little shocked. 'You know, in Australia, people often have cats as pets, and I reckon they'd be, um, surprised that you eat them. They really love their cats.'

'Oh,' he replied. 'I love cats too. I've got a pet cat. Why would Australians be surprised?' he went on. 'Don't you have fish as pets, and still eat fish?'

It was hard to argue with my guide's logic, even if goldfish *are* a different species from flathead or snapper. And I think of what he said when I try to reconcile the conversation at the market about who is entitled to call themselves an animal lover.

The answer is all in our heads. We choose some animals as meat, some as pets, and most urban folk in Australia have a clear distinction between the two. But those in the animal industry can love animals, and eat meat from those same animals, in the same way my Chinese guide could. It's all about attachment, our emotional commitment to each species, and to each individual within a species. What it means in practice is that in countries like Australia we are more likely to fetishise animals. This dichotomy, to my Chinese guide, and increasingly to me, comes across as simply weird.

What does it say about us as a nation that we are happy to treat dogs as virtual children – breeding pooches so small they can fit in your carry-on luggage, feeding them treats more expensive than smoked salmon, sending them to doggy beauty parlours – while the beef we eat has spent half its life bored rigid in a stinking hot feedlot in central Queensland? (Though the cow might, thanks to 'customer expectations', have some access to shade these days.)

What does it say about us if we're happy for sows, mother pigs, to live in a farrowing crate – a structure that need only be a centimetre longer than her, front and back – where she can't move forwards or backwards in any meaningful way, and in which she can only stand or lie down for 28 days while she gives birth and suckles her young? *For 28 days!* If you did it to a dog, there'd be an outcry.

Why is a dog – whose ancestry you can now find out through a DNA test like the one you can get done for yourself – more important than a pig?

Why is it not okay to eat horse, when every retired horse, every failed pony, must end up somewhere, and it's okay to eat cow? They're both large, soft-eyed mammals that have been domesticated for several thousand years. The butcher who first started selling horse meat in Perth in 2010 received death threats, despite the fact that horse meat is traditional fare in his native Italy. What's more, a horse is no more conscious, no more sentient – having the ability to feel pain and pleasure and fear – than a sheep or a chicken.

So our attitude to animals, as a society, is fraught. It's contradictory, and it's not helpful. Treating one domesticated animal as if it's your own child, while buying pet food for it – in the form of chicken necks from an animal that has led a paltry, unfulfilled, and very short life – is not unusual in our society. Allowing your dog to sleep on your bed, and paying thousands of dollars in vet bills when it is injured chasing down a car, while not caring about the life of your next meal, is kind of odd.

As Anna Krien points out in a particularly excellent *Quarterly Essay,* 'Us & Them', in 2012, we are becoming polarised in the way we associate with animals. On the one hand, we fetishise them – buy Dentastix to clean our dogs' teeth, watch videos of cute kittens on YouTube, believe the talking pig in *Babe* sees the world exactly as a human does – and on the other hand we have intensified animal production to such an extent that some livestock never see the sun. Some animals never feel the wind on their faces, or dirt under their feet. Everything about their existence in a so-called Concentrated Animal Feeding Operation (CAFO for short, though most lay people use the term 'factory farm') is so far removed from their genetic forebears that it's a strange testament to the ingenuity of humans that we can keep those animals alive, and relatively healthy (if possibly bored stupid), under such conditions.

Australian novelist Charlotte Wood has written about the dichotomy of our relationship with animals in the story 'This dog is not a human being ... (right?)' in *Good Weekend* magazine in 2011. In it, she recounts her confusion when a dog was awarded a war medal for bravery as though it had actually chosen to go to war and sniff out bombs. She's uncomfortable with the way 'our culture is drenched in anthropomorphic sludge'. She's suggesting (and here I'll use Krien's paraphrase), that 'the more we sentimentalise animals on the one hand, the more we brutalise them on the other'.

Wood says we have made a 'grossly sentimental failure to embrace the "other-ness" of animals ... to imagine them as anything but approximations of ourselves'. So we either imagine our domestic pets (or animals in children's stories) to be just like us – or discount any similarities at all, and accord them very few rights at all.

It's either Old MacDonald's Farm, or something to eat that didn't have a soul, wrapped in plastic from a supermarket.

Human relationships with other animals are no easy thing to navigate. It is thought we first domesticated dogs about 30,000 years ago, long before we domesticated other species – about 10–20,000 years before we domesticated pigs and sheep. At first, it is surmised, we chose the wild dogs, the wolves of the era, because they liked hanging around us, and they could be useful in the hunt. Then we realised they were social animals, thinking of us as part of their pack. We kept dogs around because we liked them, and because they were useful. Rarely is there evidence that those early dogs were eaten.

Humans, since those times, have been hanging around with non-human animals. We've socialised with them. We've bonded with them. We've formed relationships with other animals, even when intending to use them for our purposes – and, yes, sometimes that purpose was for meat. I suspect it's hardwired into us that hanging around with non-human animals is a good thing. Managing livestock and owning a dog certainly makes me feel better about the world, and there's a heap of pet owners and research to back this up.

People live longer and better lives, on average, if they have a pet. According to the RSPCA, some 38 per cent of Australian households own a dog, and 29 per cent own a cat. There are about as many pets in Australian homes as there are people – about 24 million in a recent count. And yet there are a whole bunch of animal lovers out there who still eat meat.

Okay, so you love animals?

Probably most of us do – even those who might work in slaughterhouses. Only a creepy, tiny percentage of people hate animals. Lots of people aren't fussed either way with pets, but can still admire wildlife, either in person or on a David Attenborough show. There are many who would identify as animal lovers who don't own a pet (and their reasoning could well be that owning a pet isn't always in the animal's best interests, only the human's). Some people in New South Wales would rather brumbies, wild horses, die of starvation out of sight, than condone aerial shooting of them by rangers, where the decaying bodies are more likely to be seen. It's not about a love of animals, but rather our attachments to them.

Zookeepers can be animal lovers. Farmers can be animal lovers. It isn't absolutely necessary, but on some farms it helps. The very term 'animal husbandry' implies what we as humans are trying to do – tending, and nurturing, livestock. The Old English word for 'husband', which comes from the Old Norse for 'householder', used to mean a 'farmer' – from a time when holding house and tending the land were much the same occupation. It's a branch of agriculture that involves breeding and caring for animals. At its heart, animal husbandry is about giving the animal what the animal wants, in order for us to get what *we* want (meat, eggs, milk, wool) – or at least it should be. It's about breeding and growing animals for human use, but cruelty is not an intrinsic part of this. In fact, the opposite is true. While often we humans are misguided, historically not understanding what the animal actually wants and needs, most farmers are looking to keep their livestock free of disease, free of injury. They want their animals to have low-stress lives. It's in the farmer's interests to have animals that have plenty to eat, are healthy and grow well.

When humans first domesticated livestock, we did so because we found animals that were happy to be around us, that didn't eat us, and that we had a use for. For a while we used them in our nomadic state, the animals able to move themselves from place to place as the seasons changed, or climate altered. And then, about 10,000 years ago, when the first humans did stop roaming, the animals stayed close, through shepherding, or having their own wants met. We helped feed them, cared for them when sick, watched over their young to prevent them being taken by predators. We bred them for milk, for wool, to become even more domesticated in nature, and favoured all the attributes that we liked. And when the time came, we slaughtered them and ate them.

For a long time, animal welfare has always been framed by what humans want. And what we want isn't always in the animals' interests – certainly, being eaten isn't! But can we do better when we try to give animals in our care a more fulfilled existence, a better life while in our care?

For a long while animal husbandry, the care of animals, was judged – in a farming sense – merely by how big an animal grew in a certain time. Or how much milk she gave, or how fine his fleece was, or how many piglets she had. Farming, and animal welfare, was judged by a sole criterion: weight gain, or yield, or output. The bottom line. You can tell a happy animal, they'd say (and I've heard this so often even while writing this book), because an unhappy one, a stressed one, an unhealthy one, won't put on weight.

But this reductive thinking lets animals down. We know you can get better yields by not letting animals move. We know you can get bigger pigs if you give them antibiotics. We know you can get faster-growing beef cattle (albeit tougher and drier) if you give them growth hormones. We know that if you use a genetically modified hormone, recombinant bovine somatotropin, on dairy cows they produce more milk. We can trick animals into giving us the things we want, without giving them what *they* want. We are clever enough, we have an armoury of chemicals so diverse, we understand their metabolism so well that we can drive up our yields, and drive down prices, while giving the farm animals no respect for their other needs.

Later in the book I'll look at the complexity of how we should see animals. They aren't humans. Nor are they approximations of humans. Each species, and each individual within a species, has its own wants and needs. It's up to us – as those ultimately responsible for their lives – to understand as best we can the otherness of the animals in our care. It's also our responsibility to look after domestic animals until the time of their death. That's the deal we've made with ourselves (not some contract with them, as some argue), for the pleasure of having them around.

It's also our responsibility to consider the greater environment and the animals it supports. And for that, we need to talk about cats.

CHAPTER 7

Cats. Beyond the Cute Videos

The human view on suffering is multi-faceted, but one view is that by rearing and eating animals, or using their wool or milk, we humans are causing unnecessary suffering.

Using this view, one many share, to drink milk is to cause an animal distress: perhaps to the calf and mother when separated so that you can get the milk, and definitely to the calf when it is killed – as most boy calves are, quite early on (more about this in Chapter 18). Rearing an animal for food means you are exploiting it unnecessarily, and in the process cause unnecessary suffering. So, don't drink milk, don't eat meat, don't use leather or wool, and fewer animals will suffer.

The problem is, most suffering, at least on our farm, happens in the wild. And most suffering in Australia happens not because of farming, but because of cats, and our lack of action in controlling both domestic and feral cats – actions that are within our grasp.

Feral cats are a menace in the Australian bush, while domestic moggies will also predate (stalk, catch, kill and not always eat) anything they can find that doesn't attack them. As an introduced species, cats are playing havoc with our native wildlife.

And what are vegans and animal rights groups doing about it? Not much, from what I can see.

First, let's look the problem.

It's estimated that there are at least 6 million feral cats in this country, in every landscape, and on every corner, apart from a few islands. Some estimates suggest up to 20 million feral cats inhabit Australia, but let's use the lower number, 6 million. Savage hunters, cats prey on anything smaller than themselves, often playing with the animals as they die. Two-thirds of the mammal extinctions that have already hit Australia have come directly from the impact of feral cats, according to the Threatened Species Commissioner. Currently, another 124 species, including reptiles and birds, are threatened with extinction. Cats helped send the desert bandicoot extinct, along with the rusty numbat and the lesser bilby.

If, as estimated, Australia has 6 million feral cats, and each feral cat kills up to an estimated 1000 native animals every year, the death and suffering of 6 billion animals each year really is a problem worth addressing. Directly attributable to feral cats (and hence the humans that brought them and allow them to thrive), this number of creatures killed each year dwarfs the number of livestock in farmers' care.

While the total death toll of 6 billion wild animals a year is of course an estimate, this is how the stats stack up. According to the CSIRO, Australia's pre-eminent science organisation, feral cats – the ones that have found a home in our wild places – kill 466 million reptiles a year. The CSIRO estimates that a further 130 million reptiles are killed each year by feral cats in highly modified landscapes (farms and cities and the like), and another 53 million reptiles by pet cats: domestic moggies. All that adds up to 649 million reptiles killed each year – about the same number of chickens that live in sheds for the meat trade.

According to the CSIRO, cats predate 258 of Australia's reptile species – roughly a quarter of our reptiles, including 11 threatened species. Those species threatened include some blue tongues, skinks, geckos, even snakes.

Every day, cats also kill about 1 million small mammals and marsupials across the nation – everything from introduced mice to hopping mice, young bettongs, bandicoots and the like.

These are conservative estimates according to some. The Australian Wildlife Conservancy reckons that feral cats kill about 75 million native animals every night.

Pretty bad, eh? And as I mentioned, cats generally play with their prey, as you'll be aware if you've ever seen a cat toy with a ball of wool or a mouse that it's found; that's an indicator of their nature. Death at the hands of a cat is no painless experience free from fear. It's a torturous death at the claws of a well-adapted and merciless killer.

I'm not done yet. The latest research suggests that along with the reptiles, mice and native marsupials, cats also kill 1 million birds every day in Australia. One million per day.

I'm still not done. The carnage goes beyond small animals and birds. Rangers in central Australia found the remains of a 5 kilogram rock wallaby in the gut of a feral cat – so it's not only crickets, mice and birds that are being gobbled up, but animals even larger than a domestic cat.

Feral cats also spread toxoplasmosis, a disease that causes stillborn young, birth defects and fatalities in native animals and in humans. Dispersed through the cat's poo, toxoplasmosis is particularly harmful to marsupials, and can send wallabies and bandicoots and people blind, too. In fact, cats are the only known host of the disease, and it can kill an eastern barred bandicoot within 2–3 weeks of infection. The disease also kills koalas and wombats. This is a hidden impact of feral cats.

The sheer numbers of native mammals, bird and reptiles killed by feral cats, both by predation and disease, stunned me. So I set about trying to find who in the animal welfare community is doing something about it. The answer, from my research, is nobody.

Yes, the federal government has launched a program to cull a third of feral cats from the environment. Yes, farmers and rangers I know are doing a great job in attempting to rid the world of these pests.

But what about those who run welfare organisations that aim to reduce the suffering of animals? Their voices are curiously absent. So perhaps it's not *all* suffering that matters? It's not the *total* suffering that matters? Just the visible suffering directly caused by industry? Maybe some forms of suffering are just more apparent to them, and that's the suffering they don't like, or don't want to countenance?

High-profile animal rights campaigner Brigitte Bardot has criticised the Australian federal government's action to cull cats. PETA, the People for the Ethical Treatment of Animals, objects to Australia's feral cat cull because the cats might suffer. (They also suggest that one of the ways to deal with the feral cat problem is to become vegan, though what the connection is, I'm not sure.) But no thought, it seems, is given to the 1000 or so animals each and every one of these cats will kill each year. If reducing suffering was the aim, wouldn't drastically reducing feral cat numbers be one of the first places to start?

So who else is out there doing something to reduce feral cat numbers? I was disappointed, but perhaps not surprised, that there are some remarkably small voices when it comes to standing up for *all* animals.

According to their website, Voiceless is opposed to hunting or poisoning any animal (including 'feral' cats) for the sake of 'conservation' (their inverted commas). They believe these issues 'can be successfully addressed in a manner that does not cause suffering to animals. This may include fertility control, fencing or other non-lethal means of population management.' But this statement does nothing to counter the wholesale slaughter and compromising of our precious native wildlife that is happening each night at the paws of rogue cats across the continent.

In response to articles such as one in February 2017 in *The Sydney Morning Herald* that talks about feral cats and the chance to cull them and protect native wildlife, Animals Australia comes across, sadly, as equally ambivalent. It said it was sceptical the government's proposed measures – including $5 million of funding to cull feral cats – would make a significant difference to the wild cat population.

Does Animals Australia have an alternative plan that would reduce the wild cat population? At least Bardot and PETA suggest sterilisation as a starting point. Instead, Animals Australia has posted a rant on its website about how feral cats have the same rights as pet cats – which in turn should have the same rights as people. The essay's author, Frankie Seymour, was once the Animal Justice Party candidate for the Federal House of Representatives in the seat of Eden–Monaro. It seems only some animals are the subject of justice – just not the victims of the feral cats that blight our nation.

Other organisations aren't much more helpful on the topic. On the Australian Wildlife Protection Council's website, for instance, there's little mention of feral cats and the 1 million or so birds they torture and kill every day. It seems mostly interested in large, cute, warm-blooded, furry animals, like kangaroos and Tassie devils, with barely a mention of small marsupials, lizards and the like. Admittedly they're first and foremost an anti kangaroo-cull organisation, but surely native animal protection (and suffering) is more complex than this organisation makes out? They are, after all, called the Australian Wildlife Protection Council.

The Cat Protection Society of NSW, a member organisation of Animals Australia, does good work to disseminate information on why everybody should de-sex domestic cats, and in encouraging accountable cat ownership, but even they take no responsibility for reducing the actual damage feral cats do to our native wildlife, and the suffering that goes along with that damage. Yes, the cousins of those cute, aloof, inspiring moggies that grace so many greeting cards are free to do untold killing every day, thanks in part to the lack of leadership from an animal-focused organisation.

It saddens me, as someone who sees with my own eyes the impact of toxoplasmosis on native wildlife, who has bandicoots and birds and lizards predated upon by feral cats on our farm, that so many animal-focused groups seem incapable of having an adult discussion on cats: their impacts, their torturing of other animals, their ability to send other more vulnerable species extinct. The rights of the cats seem to take precedence over the rights of millions of native animals killed each day, and I do find this attitude smacks of something close to the hypocrisy such animal rights groups object to every day. Such groups think it's hypocritical to say you care about animals and use them for food, while at the same time claiming they care about animal suffering but do little to alleviate the greatest cause of animal suffering in the nation.

At least the Australian Wildlife Conservancy is really doing something constructive. They're conducting the largest feral cat research program in the nation's history, to try to work out the best way to deal with feral moggies. They're monitoring cats, tracking cats, trialling fenced areas, managing landscapes to provide more shelter and protection for native wildlife, all the while trying to understand the best way to eradicate cats – and all the while being fairly under-resourced. My question is, when it comes to dealing with such a dangerous introduced pest, where are the other animal welfare groups?

I write this not to condemn, because we can all only care about so many things. For some it's refugees. For some it's the colour of their eye shadow, where they take their coffee, perhaps, which charity they support, or the origin of their steak. But in dealing with the feral cat problem, I do wonder about the dearth of support – and on occasion antagonism – from animal welfare groups.

Of course, feral cats are only one predator humans have unleashed on our fragile ecosystem; I've not even looked at foxes or wild dogs. But I wanted to bring up cats because of their inherent cruelty (their desire to play with prey, to kill it, and not use it for food), their revered place in our society, and to show that dealing with suffering isn't simple. It goes to the heart of how complex this topic is. From the perspective of the health of our native wildlife, it is hard to see an organisation as taking their animal welfare responsibilities seriously if they ignore the devastating impact of the humble moggy.

———·———

It is possible, with some effort, to reduce the damage wreaked by feral cats substantially, and in the process reduce the total suffering inflicted on so many other sentient beings.

Suggesting that the feral cat that I will trap tonight has as much right to this land, while it torments and then slaughters blue tongues and bandicoots and hopping mice, is just bunk. I might remove the feral cat's right to live, but in the process I hope to decrease the suffering that happens on my land, and also help maintain its indigenous animal population.

I will kill any feral cat I trap, and we have regular incursions from feral cats on our property. And I will defend this practice to anyone who asks, because of the overall reduction in suffering this brings, and for the protection of native wildlife.

I will argue my point that suffering is suffering, and that native animals aren't adapted to withstand the onslaught of feral cats, to any who care to listen.

I think it's a shame that all human-induced suffering isn't considered equal. You can defend a feral cat's right to live – but I don't think you should do it under the guise of 'animal welfare' and expect me to remain silent.

We are conflicted in the way we look at animals, as my visits to intensive farms has shown. I know it's so, and I accept it's so. Just *why* it is so is the subject of the next chapter, where we look up close and personal at what farming is, what it does, and why it's hard to argue who has the right to moral superiority.

CHAPTER 8

A Farmer's Musing on Suffering

I was driving home from a family outing one day and saw a hawk on the road ahead. It looked like it was attacking a rat. We were driving fairly slowly on a dirt road that was more like a single-lane track. People don't, and can't, drive very fast there – and yet most mornings you'll find a dead wallaby or two on this stretch, a sign of a healthy wild population, but a confronting sight nonetheless. Tasmania is, after all, considered the 'Roadkill Capital of the World'. Estimates put the native roadkill toll in Tasmania at 300,000 animals a year, and partly it's because we have *soooo* much wildlife.

Anyway, on this day, as we got closer, I saw the hawk fly off – and saw it was no rat that it had been attacking. It was a reasonably nude, virtually hairless baby wallaby. A velvet, they're called, because of the texture of their skin. Its mum lay dead to one side of the road, struck by a car in the night, presumably. The baby, finding no milk from its mother, had ventured for possibly the first time from its mother's pouch. Out in the open, it was easy prey for the hawk. By the time we found it, the hawk had pecked both its eyes out and it was bleeding from the mouth. But it was still alive.

Faced with both a human-caused dilemma (the mother struck by a car), and nature doing what nature does (a raptor having a crack at an unprotected young wallaby, which they do all the time in the wild), I felt paralysed. Should I intervene, because the baby was only orphaned through the actions of humans? (In the ideal world, those who hit a marsupial with their car are responsible for checking the pouch and taking orphaned young to emergency wildlife carers.) Or should I now let nature take its course and let the hawk have a feed?

I chose to intervene. Deciding in the moment that a blind baby wallaby could probably not be rescued by a wildlife carer, I chose to end its suffering. I clonked the fragile marsupial on the back of the head, checked it was dead, then shifted its body, and that of its mum, off the road. I moved the body and drove off.

Our then five-year-old son, Hedley, was in the back seat. And he was livid. He was so angry with me for taking the life of this joey. He'd been to the local wildlife rescue sanctuary, and one of his teacher's aides was always bottle-feeding wallaby young, eternally carrying joeys in a pouch on her chest, rubbing their genitals with a cotton bud to promote their excreta (nature kindly toilet trains marsupials to respond to this so they don't excrete urine or faeces into the pouch). Hedley berated me for taking this life, and still remembers it, five years later.

Should I have scooped the blind baby wallaby up and taken it to a wildlife carer? I doubt it would have survived. But that's what wildlife carers do: against all the odds they nurse native animals injured on the roads. I can't know exactly how that baby wallaby would have coped if we'd scooped it up and sent it to a carer, but I really doubt it would've survived.

Farms are different. If a farm animal was damaged in such a serious way, at such a young age, we would be required to end its suffering – required by our own standards emotionally, and perhaps legally. Imagine if it was a young piglet, born pre-term and just as vulnerable as the wallaby, that had been attacked by a hawk after its mother had died? What would I do then? Or an orphaned lamb, bleeding from the mouth and eyes, unable to stand or run? Wouldn't I have felt the need to end its suffering in the same way?

—·—

I want to tell you a little about our farm, because it does frame the way I think, and perhaps helps to illustrate how growing food makes you view the world differently.

Our property takes up the floor of a narrow valley about an hour south of Hobart, in Australia's southernmost shire. Fat Pig Farm is a 70 acre (28 hectare) mixed farm where we live and work. About 55 acres is usable for agriculture. The rest is reserved for wildlife, a combination of regrowth bush with wattles dominating, older bush where tall, handsome gums stand sentinel, and a winter creek where a mix of introduced and native species compete for space and light.

On any day we'll see wallabies, bandicoots, bettongs, quolls, eagles and hawks, along with ducks and swans and native hens, tiny wrens, finches and fantails, squawking ravens and carolling thrush. The flame robins vanish over summer, and reappear about the time of the first frost. Black cockatoos make guest visits to strip the bark from the silver wattles and crack open the radiata pine cones, before screeching off to find better food elsewhere. Skinks scatter around in sheltered nooks, frogs croak loudly in the night, drowned out, at times, by the fiendish sound of devils, shrieking in the bush.

I want to talk about the farm, the place we call home, because it has informed a lot of my thinking about meat. We've made mistakes, we continue to make mistakes, and we're no prize example for all to follow. But I want to talk about the place because it has helped me to see the issues around meat in their true, more complex form.

Fat Pig Farm is broken into about 15 distinct paddocks, where we graze livestock, including our namesake pigs. We also run a sizeable market garden, with about 1 kilometre worth of annual garden beds, along with about 40 fruit and nut trees and several rows of perennial berries and the like. The farm is minimal input (meaning we don't buy in lots of fertiliser, and try to grow as much of the food for our animals as we can, and so on), while the 2 acres of market garden is run on organic principles. We are in service to the soil, and we try to build a complex ecosystem below and above that narrow crust of dirt that rims our little bit of the earth.

Despite the fact that we farm foreign plants and animals native to distant climes, we're trying to build a new kind of harmony on our land. It's our aim to try to let nature thrive, to let our livestock express their instincts, and provide space for both these introduced species and the native animals and plants to live unhindered. So on any one night we might have 300–400 wallabies come onto the wildlife corridor to graze. We might see several possums meandering around the place, too, along with smaller marsupials, including the predators – the quolls and devils and owls. It's hardly a pristine wilderness, but there is wildness within our boundaries.

At any particular moment in time we're responsible for a whole bunch of lives. We have three milking cows and their calves, and a small beef herd numbering up to ten. As I write there are 11 goats, 28 pigs (including four breeders), eight chooks, several thousand domesticated bees, and we also have some duck eggs in the incubator. On our farm, we're responsible for the breeding, health and wellbeing of our domesticated animals. If a pig gets sick, we care for it, calling the vet as need be, administering medicine if required, isolating the animal to rest if injured. If there's trouble at birth, we intervene. Where possible, we let nature take its course, but only insert ourselves in the process where it looks like nature might go awry, as humans have done for thousands of years. Animal husbandry, caring for livestock, is older than agriculture itself.

We take this responsibility seriously. A snake-bitten pig? We do our best to save it and make it comfortable. A calf struggling to suckle? We try to teach it to latch onto its mum, often putting ourselves in harm's way. A ewe distressed while trying to give birth? We get in and help pull the lamb so the mother doesn't tire herself to death. We also have to try to work out when there is no hope, and then end the pain as quickly as we can. If anything, we let animals live too long, fending off the results of illness or injury, only to extend their suffering. Knowing when all is lost is an art, one we are still only starting to learn. In all of this, we are not alone; most farmers are very aware of their livestock and its needs.

But most of the suffering on our farm of 50-odd domesticated animals isn't anything to do with us. Most of the suffering on this farm – and most of the suffering of animals around the nation – happens because nature is cruel. Animals get old or sick routinely. Wallabies get listeria or toxoplasmosis, or simply get old and start to hop lamely around, even during the day. Birds age and get hurt and can't fly well. And all the while the predators wait. Crows, black as pitch and ever watchful, can sense a weak animal before it falls. The eagles and hawks are acutely aware of a young animal separated from its parents. At night devils and quolls – opportunistic predators who would rather eat carrion, meat from the already dead – won't hesitate to strike if something weak or vulnerable crosses their path.

Watching nature do its thing has taught me a lot. It has also raised more questions than I can answer.

What is suffering? Is it the pain and fear I imagine a lame wallaby has in its last days, just waiting until something stronger and bolder takes it down? Is it only suffering if it is seen and then ignored by me? All day, all night, animals predate on each other. All living beings, besides plants and algae, have to eat other living things. Some animals eat plants, some eat meat as well as plants, and some eat virtually only meat. In this daily struggle some beasts are always prey, their entire being ever vigilant against the threat of being eaten. All living creatures have survival strategies, but some can only run or fly faster than the animal next to them, each minute of their lives a lesson in awareness and anticipation.

Why is it that I'm only responsible for the immediate suffering I see, when most of the suffering on this farm happens just out of view? It happens at night, in the long grass, on the fringes of the forest. It happens behind the trees when a goshawk has a clear run at a potoroo. We witness the daily diminishing of a flock of wild ducklings, whose mother obviously can't protect them from attack as they sleep at night. I find the feathers and fur of innumerable animals over the weeks as I traipse the farm each morning. Very few would have died as quick as one of our farm animals. I'll find half a frog, the gnawed leg of a joey, the head of a finch: grim proof of nature at work.

None of the wildlife that lives on our farm has their injuries cared for, their babies watched over by another species, unless that other species had hunger on their mind. I don't tell you this for any other reason than it makes me think.

Animal 'suffering' is a very human construct. Other animals really don't care how much suffering they inflict; they just do what comes naturally. Humans, for some reason (and it could well due to be our ability to 'reason'), see suffering, and many of us want to stop it. We project our own parenting instincts, our own ability to feel loss, our own sense of pain, onto the animals around us. But it is only with the animals that fall into our immediate care, or the ones we have forged some attachment to, whose suffering becomes a part of us. Most of the time, nature inflicts pain, loss, hardship and fear with no accountability, no one to blame. A hopping mouse taken by a quoll? It suffers as much or probably more in its last minutes than a chicken raised in an intensive shed that is gassed at the end. A rabbit kitten, struck then swallowed by a snake? Probably a bad way to go, and not as fast as a hunter's bullet.

But for some reason, and reasons I probably agree with, a rabbit actually farmed on our property as opposed to born in the wild should expect a healthy life and a quick, painless death. The rabbit that the snake suffocates and eats? Well, it's not my role to get involved.

It would appear that our idea of animal 'suffering', which would be considered normal in the wild, is simply a modern intellectual construct – one that we have the brain power, the luxury, and the time to frame. I think this notion of suffering, of what it means, of what our role is, came home to me when the baby wallaby was in front of my eyes, suffering, a few years ago. And, of course, it was a child's eyes that brought it into full view.

———·———

In Tasmania, wildlife thrives. Over a million animals, mostly wallabies, are killed each year to protect farmers' livelihoods. I have a crop protection permit for our farm. I'm allowed to kill wallabies and brush-tailed possums to protect our pasture and gardens. If I kill an animal, I'm required to check the pouch of the females and knock any babies on the head, because animal carers are already overwhelmed with joeys orphaned after wallabies and roos are hit by cars, and can't cope with the number of extra orphans farmers could potentially bring in. It's one rule for on the road, another for the farm. Same animal, different potential fate, same suffering.

For me, this puts the farmer's view in stark relief. Farmers know animals live and die all around us. We know we sometimes have a hand in that, and sometimes *should* have a hand in that. We know that growing vegetables is a reason to kill marauding wildlife. Consumers, after all, want their greens without holes, their vegetables cheap, their corn not nibbled.

———.———

We try to prevent unnecessary suffering on our farm. And by unnecessary, I mean a certain degree of pain or fear in livestock is tolerable. Animals experience discomfort, presumably, when they give birth, just as humans do. And also presumably they have the hormonal benefits to cope with that pain. Animals are also often frightened. Livestock are mostly prey animals, used to being cautious, if not downright scared; this is their normal way of being, not something we've foisted upon them. They can also suffer the normal injuries we humans experience with an active lifestyle. Abrasions, sore feet, mild illness – these are to be expected. Where these become abnormal, or too common, or problematic, that's when we get involved. A life free of pain and fear is an impossibility, and so it's partly a matter of point of view as to what is okay for an animal to feel before we insert ourselves in the situation.

Piglets on our farm are sometimes crushed by their mothers. It's the sad reality that in their efforts to get close to mum, to suckle, they can be smothered by her 150 kilo bulk. I find this heartbreaking. Always have, maybe always will. But it's the reality that even wild piglets get crushed by their mother, despite having far smaller litters than farmed pigs. It's also the reality that pigs have multiple young simply *because* they lose piglets. The small, vulnerable babies are easy prey for wild dogs, for hawks and eagles, for predators of any size.

So we know that suffering and pain, fear and loss happen. We know we are just cogs in nature's wheel, and that what happens on our farm is only partly in our control. We want our livestock to experience a certain instinctual existence, but all the while caring more for them than if they were actually wild.

The problem is, we often fail. It's part of the reality of human existence, of our limitations, our very nature. I call it the supermarket test.

The supermarket test is easy to conduct. When researchers ask consumers to name what's important to them in regards to animal welfare, people often cite the ability to roam, the ability to express their instincts. Only a few would be able to look at the footage of battery hens and say they really want to support that system. The problem comes not with our intentions, because most people have good intentions. The problem comes with our actions. Because no matter what we say about sustainability, about ethics, about our desire not to harm the animal, the environment, the world we live in simply with our food choices, when faced with items on the supermarket shelf, we buy the cheapest.

We say we hate the idea of battery hens, then buy cage eggs. We are shocked at seeing pigs in sow stalls for virtually all their adult lives, but will happily buy the cheapest pork chops we can, knowing (or perhaps suppressing) in our hearts that the system they came from wasn't as good as the one that produced those expensive chops from an organic farm. We have our standards that we are proud to stand by, and then fail our own standards when it comes to the reality of what we put in our shopping basket. We all believe that we're better than we probably are. According to a 2010 Newspoll, 'A Pound of Flesh', conducted for the Vegetarian/Vegan Society of

Queensland, 99 per cent of Australians are against cruelty to animals. According to Vegetarian Victoria, the 98 per cent of us who do eat meat 'eat animals and fish that come from cruel, intensive farming practices!' There's nothing particularly unusual about this, sad though it may be to admit. It's a normal human trait – but it's good to be aware this paradox exists.

On our farm, we fail the supermarket test, just like everybody else. I am particular about my pork, unlike many – but yet I have bought down jackets that may have been the product of live-plucking of ducks; I just didn't know I had to research this before buying the jackets. I rarely eat mass-produced chicken because I have doubts about the system big farms have of rearing them, but I'll buy a work desk with no research into the environmental outcomes of its manufacture, or the conditions of the people who made it.

On our farm, we believe in soil health, and animal health, but have to constantly make compromises about what we can and can't do – in terms of our time, our resources, and the practical realities of geography. All of us are limited by our knowledge, finances, location and more.

I believe in upholding certain standards, and yet in the day to day, there are times when I don't manage to do what I set out to achieve. Animals probably do suffer on our farm, when if I had all the information, if I could be everywhere at once, they wouldn't. I strongly believe that the concept of suffering and its minimisation, while it is a human construct and it's hard to know where the boundaries of our actions lie, actually is the responsibility of all who farm, and all who use what farmers produce. In fact, it's the responsibility of all who walk this earth who are capable of higher thought, empathy and compassion.

We *are* better in our heads than we are in reality, and our aim on our farm, and one of the aims of this book, is to narrow the gap between the two.

CHAPTER 9

Floyd and What We Don't Know About Animals. Yet.

Bobby is pretty damn cute. He nuzzles me, licks me. He bounds up when I come near, answering to his name. He chases the sheep, and loves nothing better than a head or tummy rub. When he tucks his head in for me to give him a good old scratch around the jowls, however, I end up taking my life in my hands.

The reason? Bobby is a steer, a 24-month-old Jersey–Hereford cross with big horns and identity issues. Orphaned at birth, he was hand-reared by us, before spending most of his days being watched by, and watching, the dog. What he knows is that his mum looks like a human, his best friend is man's best friend, and that sheep are there for the chasing. Bobby is part cow, part dog, part human teenager – his nurturing a major factor in how he sees himself.

Getting to know and understand Bobby taught us a whole bunch of stuff about animals that we didn't really get before. Like the deep 'imprinting' of their parents on a young animal, so forever after they will see its parents as their guardian – a being that provides food and safety and shelter. It didn't matter that Bobby was only really in our immediate care for three months, then ever after with his own species. He would always see us as parents.

One of the defining characteristics of owning animals is that you'll never really understand them. This shouldn't be a shock, because many of us have lived with humans, under the same roof, sometimes with those from the same genetic stock, and struggled to understand their behaviour. So it seems odd that it has taken farming groups, and the researchers who work with them, so long to spend the money and time researching the behaviour of the non-human animals that we farm – often intensively – just to get the basics right. It took research teams until the 2000s, for example, to learn that chickens are individuals, with complex thought and social interactions.

The arrogance of humans, it seems, is that we think we understand the animals in our care, when we only know the half of it – if that.

On our farm we've had pigs for a few years now, and understand some of their intelligence and behaviour. We know that they are clever enough to find food. Put a single pig in a maze on one day, then the next day watch that pig lead their siblings into a dead end, lose them, then go and get the food itself. We know pigs are trainable like a dog. We are getting better at understanding how they view the farm, their surroundings, their predilections. And yet we still get surprised by tales from others in the game.

A friend of ours had a breeding couple of pigs and fattened a few litters over the years. Floyd the boar and Nellie the sow lived happily together, long after Nellie stopped having litters. Until one day, she got old and died. A space was found, a hole dug on the farm, and Nellie was buried deep in the earth. Floyd was living alone, and was obviously distraught. He paced the fence line, and eventually broke through to get to the grave, then proceeded to dig.

Now, to give you an idea of the scenario from the farmer's perspective, just think: a 200 kilogram boar, with massive tusks and tremendous strength, is no easy beast to move. It's no easy animal to deter from its course. And knowing pigs are omnivores, an animal that would happily eat meat, or a carcass in the wild, the owners looked on, powerless. And mortified.

Floyd dug. And dug. And dug.

He dug down so deep that Nellie's trotters appeared out of the earth. And Floyd, once two trotters were exposed, lay down with his head between them, and barely moved for a day.

He then walked off, straight back to his paddock, and never escaped again. What he was thinking, if he was grieving, or if needed to rationalise his loss, we'll never know. But it does show that there's a whole bunch of stuff we are yet to learn about the mental and emotional state of animals in our care. And stuff we'll never have the language to be able to communicate with them.

I tell you this story to illustrate a point. To show that what we know is possibly still less than what we don't know. And that we shouldn't assume that animals do or don't feel things that we feel. It would be easy to anthropomorphise this behaviour, and see Floyd as an old man who'd lost the love of his life. Or to perhaps see him as simply curious about where his former bedmate had gone. Or hungry, then finding the meat too putrid to pursue. But most of all we should look at animals in a new light, as those able to feel pain, and loss, and confusion – not just fear, or loathing, or famine.

In late 2017, we had a litter of pigs born to our sow, Audrey, one evening. In unlikely fashion, instead of staying close to mum like they normally would, the litter for some reason separated almost immediately. We found three or four wandering around, lost in the dark, only discovering them because of their little homing grunts. All up, we found nine, and plonked them back next to mum overnight to suckle and cuddle.

Problem was, there was one still wandering that we didn't find. The next day, one skinny little piglet was found meandering 100 metres away from mum. And despite multiple and repeated attempts to get it to suckle, it never latched onto a teat.

Farmers are often running mini orphanages. For animals rejected by their parents, or actually orphaned at birth, the farmhouse can become a second home. Lambs are often bottle-fed, calves bucket-reared, ducks incubated in the house. For us, this was the first baby pig (called a 'slip' in the trade) to land up at the house. At this stage, we didn't know whether the pig would make it or not, as it seemed unlikely it had suckled and garnered that all-important colostrum, the early milk full of antibodies and extra fat. Without it, often an animal will never thrive – their gut much more susceptible to bugs, their body ripe for attack from microbes.

And for a few days, while we concocted drinks from cow's colostrum (the best we had on hand, and better than no colostrum), fermented milk and vegetable oils, the pig, who we named Findora, was fine. Then a few days in, the scours started. Scours, a form of diarrhoea, are smelly, watery, foul and are often followed by death as the lining of the digestive tract is stripped, and barely any nutrients stay in the animal's body long enough to be absorbed. My partner Sadie was around all day, and up several times a night to feed the pig, the bathroom reeking of its messed-up digestion. But somehow, after about a week of looking frail, and thinner, Findora pulled through.

It was about this time that we moved Findora from the bathroom to an apple bin outside our bedroom window, which had a lid on for protection and warmth at night. Findora thrived in our yard. He soon learned how to escape the box during the day. He would follow us around, chewing our shoes, skating around the kitchen on our polished wooden floors thanks to the tiptoe walk that pigs do on their trotters. He'd nuzzle the washing out of its basket, nibble our toes, and generally squeal with delight, and hunger, at every sighting of us. He was remarkably stubborn at moving from the bottle to solid food, resisting every attempt we made. Long after his siblings outside had started to eat from a trough, Findora was adamant he only wanted a bottle. He also, despite lots of play-dates in the paddocks, never really bonded with his brothers and sisters.

At about three weeks old, one of Findora's siblings was taken by an eagle. Without company during the day, we had worried Findora might become fodder for the breeding pair of endangered wedge-tailed eagles that float on eddies of wind above our farm. We saw the eagles regularly at the time, often over the house, always on the prowl for food. With only a couple of hundred breeding pairs of these magnificent wedge-tails left in Tasmania, a lost piglet is part of the price we pay to keep the population alive. Strangely, however, it wasn't the solo pig, with no mother there to protect him that the eagles took – rather, one of the nine in the paddock with mum. A tragedy, but one where nature – our particular kind of wild Tasmanian nature – took its course.

After two months, we were ready to move Findora to a new home: a paddock with his siblings, other pigs for company. Yet, just before we moved him, one day, suddenly, I found him dead. From feisty,

89

healthy, playful and forever present, it seemed his curiosity had got the better of him, his fate sealed at the dangerous end of a passing tiger snake.

Man, how we grieved. Even though we were getting tired of the novelty of having him under our feet, sneaking into the house, nuzzling the windows, waking us up early, it felt like such a loss. Grief gripped us.

Findora was often up at the restaurant, meeting our guests. We watched his growth, his learning, saw his personality change and mature. We'd bonded with him like no other.

When he died, my partner Sadie wrote this about him:

> *Findora died late last week. I'm sadder than I thought I would be. He used to hang out with me under the clothes line: when he was little, he would sleep off his bottle in the sunshine and when he got too old for long naps he would nuzzle my boots hoping for more food. In the morning, at the sound of the door, he would peer over the edge of his hay bale house, and leap out, scattering hay.*
>
> *When we first started posting (on social media) about Findora, many of you asked whether he was intended for bacon. Surely after all those late nights of bottle-feeding he would grow old as a pet. I've thought a lot (A LOT) about this. We loved Findora because we got to know him. We named him. He had a personality. He was curious and playful and loved (LOVED) having his ears and tummy scratched. But every other pig on our farm has a personality that we haven't had the time to realise. They play with each other at dusk. And every pig in every supermarket/superfund-sponsored factory*

farm has a personality that we will never get to know.
And we, those of us who eat meat, therefore, have
no qualms about eating them. We don't eat our pets
– of course. But what if we nurtured Findora into
adolescence, loved him and gave him tummy scratches,
and then killed him (quickly) on the farm for food?
Is that better than eating anonymous meat? Meat
from an abattoir where a slaughterman has to kill an
anonymous animal every twelve seconds? Meat from
anonymous animals that are trucked across the state
and held for hours or days in yards with unfamiliar
animals and unfamiliar smells before they are killed?

Or is it better to truly care for the animals that
we eat?

I don't know. And I didn't have to make the decision
about Findora.

Written in the moment, there's a deep sadness to this that you can feel – but also an anger. An anger that Sadie felt for all those other, unnamed and unknown pigs. In it I also sense her disappointment: that the general public may not see all those pigs in sheds on intensive farms as having souls, as having personalities, like Findora did.

We found ourselves looking at our own double standards, our own weird anomalies of logic. Why did we feel this death so much, strangely way more than Findora's sister when she was taken by an eagle?

———·———

How non-human animals think has been debated for hundreds if not thousands of years. In relatively recent times, in an attempt to get some more animal-focused criteria out there, so farmers can do things that have more social licence, there's been work to ensure animals have some basic rights.

These overarching requirements of animals have been dubbed 'the five freedoms'.

The first is freedom from hunger and thirst.

The second is freedom from pain, injury and disease.

Next is freedom from discomfort, so the animals have an appropriate shelter and somewhere comfy to rest.

The fourth is freedom to express (most) normal behaviour. (Note some people specify *most*, because only some behaviour might suit domestication and our farms.)

And finally, freedom from fear and distress – essentially, avoiding mental anguish.

While these freedoms are a great starting point, used by the RSPCA and others as guiding principles, they don't really have the depth or nuance that modern animal husbandry demands. Take the following example, using stress as a guide.

Scientists measure stress in animals frequently in their research. Measuring stress hormones is one indicator used to define how we should house animals. For example, free-range chickens are more likely to show higher elevations of one of the stress hormones, corticosterone, than caged chickens. Remember, the ancestor of the modern chicken was a nervous prey animal, forever on alert for predators, both in the air and on the ground. Being alert is a fairly normal state for them. So, yes, it's likely that birds allowed to roam around may have more corticosterone – but you could also argue that's a more natural state.

I often think of the natural state when I try to consider livestock. And sometimes I consider what we humans do and if it has any relevance. In terms of stress, or freedom from getting hurt, would

I take my son, my family, and shield them from all harm by not letting them outdoors? Is isolating them from potential risks giving them the fullest of lives? My son experiences fear, anticipation, joy, trepidation, surprise and delight when he ventures into the open air. Is it okay to deprive an animal of these feelings, in their relatively short lives? Do they even have these feelings in the way we understand them? Am I anthropomorphising just by making this comparison?

———·———

When you place animals in environments that are alien to them, such as sheds and feedlots, it's harder to see how farmers are allowing for the five freedoms. The Australian Lot Feeders' Association refers to the five freedoms on its website, and has even put out videos to explain how feedlot cattle match these criteria. Of particular interest to me was the 'freedom to express normal behaviour'. Professor Andrew Fisher, from the Faculty of Veterinary and Agricultural Science at the University of Melbourne, fronts the video and explains the three things that define normal behaviour. The first is the ability to lie down. So, a bit of space. The second is the ability to socialise with other animals. Yep, a couple of hundred head in the same pen ticks that box. And the only other behaviour referred to is the ability to ruminate – to be able to chew food a second time after it has been through the first stomach, the rumen.

I'd agree with the fact it's a good thing that feedlot cattle get to express all those behaviours. But what about the myriad other behaviours that cattle may want to express? Is standing, lying, eating and having company really doing this job? Without knowing exactly what the cattle think, it's hard to know. But it's also unlikely that what – just by chance – exactly marries with the farmers' needs has also given the cattle the freedom to express behaviour that they need or want to express. In another video on the same industry website, talking about freedom from discomfort, one feedlot worker states that 'one of the things that a lot of feedlots do is build shade'. Not all feedlots, or even most feedlots, just a *lot* of feedlots build shade.

Partly, what we know about animals is tainted by the research. When I've spoken to scientists, who are usually funded at least in part by the animal industry that they are researching, there's been reticence to address big issues. Such as global warming when you're talking to researchers about virtual cattle fencing (where cows are given a collar that lets them know, via a small electrical current, when they're overstepping a boundary set via satellite). Or sow stalls when you're discussing a pig's ability to have higher thought, and navigate using a mirror. Scientists self-censor, obviously aware of who pays the bills. It seems only fair that the industry that rears the livestock should pay for some of the research, but in reality there's the perception – and I've seen that reality – of results meeting industry needs, not animal needs. Truly independent research is needed, not only to verify industry results, but also to look at what is actually best practice, rather than just creating band-aid solutions to existing problems.

Research that is run by the industry takes the status quo as the norm. Research that is independent has the ability to look at all options and find the best solutions.

To give you an idea of how that could look, take farrowing crates. These are restraining cages that lock a sow in while she gives birth and suckles her young. The mother can't move in any substantial way except to lie down. The cage is built of solid bars, or sometimes with panels. It has to be at least 1 centimetre longer than the sow, and 1 centimetre wider than her, so she isn't necessarily touching it all the time, though for practical reasons it's usually a bit wider than that, allowing the mother to wriggle into a lying pose, but not allowing her enough room to take step forward or to turn around. Below the bottom metal tubing there's room for the piglets to run in and out, and so while the sow lies down the piglets suckle. Essentially, it's a sow-sized cage. The idea behind their original invention was to protect the young piglets from crushing when the sow lies down. The piglets can still get crushed under her as she lies, but there's a better chance of them having time, and space, to slip away. The mother pig is put into this cage a few days before giving birth (farrowing), and kept there for about four weeks – namely, during the birth, and for about three weeks or so after. So, while Australian Pork Limited is trying to rebrand the farrowing crate as a 'piglet protection pen', others have been researching ways around the crushing problem.

And research has found that new designs, where there's the option to restrict the sow's movement just for the first week, then open the pen up more so she can move around for the following two or three weeks, are gaining some favour. 'Farmers who have made the change to the new farrowing pen designs have not reported any economic disadvantage aside from the initial cost of installation,' it says in the latest *NSW Pork Industry Overview* report from the NSW Department of Primary Industries.

But imagine if you looked at all the options to protect against piglet crushing, rather than retrofitting sheds? Imagine if research was conducted into breeding for mothering instinct, or a diversity of systems, rather than improvements on the 'protection pen'? Would we come up with better results, in the long run? Are the people who work essentially for the pork industry truly able to criticise the practices of those who pay their bills, and still get work? It seems unlikely. Research funding is fickle. It's often political. And more and more it's been made quite clear that even groups like our once lauded CSIRO must have more industry funding to ensure their survival. What suffers is pure research, and what really suffers are animals, because the best solution for them may not suit the industry that determines the scope of the studies.

For all the Floyd stories that must be out there, there's relatively little novel research being done into how animals actually think, as opposed to how we can manipulate them on the farm.

For the foreseeable future, despite advances in our understanding, animals will be largely unknown and unknowable. If we think we know it all, we're wrong.

Animals have so far defied our attempts to unravel how they feel, what they think, how they perceive the world. That's partly why we like being around them – because unlike other humans, not knowing how your dog is feeling, except it's happy to see you, simplifies things. Presuming your cat has only its self-interest at heart frees us up for other concerns. Not having to worry about whether you've praised your pet ferret enough gives us a kind of freedom we otherwise might not enjoy.

But realising that animals are different – that they are, simply other; that they aren't human – should give us cause to at least try to give them a life that has meaning in their terms. A life where they can express behaviours.

Behaviours like sows nesting, chickens dust-bathing, ducks swimming – these aren't a long stretch of the imagination, but they do seem to be a long stretch for the intensive animal industry in Australia to contemplate.

And it's up to us to ask them to change.

CHAPTER 10

It Ain't Pretty:
What It's Like Being at a Kill

I change.

I stiffen. I tense up on days when we're going to conduct a home kill. I brace myself for hours before the animal dies. If I was the beast to be slaughtered, my muscles would be dark, my meat tough. The adrenaline coursing through my body alters the chemical composition of my muscles. Coincidentally, because of the chemical changes, my bacterial count would be higher than normal if I was the one going to the gallows, yet this animal before me doesn't realise its own death is coming.

There is nothing, I repeat *nothing*, nice about seeing a warm-blooded animal take its last breath. Anybody who thinks there is probably shouldn't be anywhere near a kill.

We dispatch animals regularly on our farm. We do it at home because it's kinder to the animal, not kinder to us. In later chapters I will discuss why we think it's okay to eat meat. In fact, I believe it's no less cruel, no less harmful than eating vegetables – at least the way we rear our animals, on our particular patch in Tasmania's Huon Valley, from what I have observed from my time on the land.

But trust me when we say we take no pleasure in the deaths that happen on our farm – neither those forced on us, nor those we choose to take.

When I say forced on us, and as I mentioned earlier, there's a saying in the farming industry that if you have livestock, you have dead stock. It's a fact of farming life. More than that, it's a fact of life generally.

Animals live and die at our command – but most often on our farm, they live and die *not* at our command. That may not be true of all farms, but it is here. Why? Because we have a healthy, if remnant, ecosystem on the fringes of our land and through the centre. Quolls, those magnificent tiny marsupial carnivores, thrive here in both the brown and black forms. Bandicoots gambol in the paddocks overnight. In fact, most of our ground-dwelling wildlife are only visible at night. We welcome them, even if they compete with us, and the animals in our care, for food. Brush-tailed and ring-tailed possums. (The brush-tails have become more a ground dweller as their natural predators, the thylacine, and to some extent the Tasmanian devil, have vanished.) Antechinus, brown bandicoots, long-nosed potoroos, Bennett's wallabies, pademelons, even the endangered Tasmanian devil parade around the house at night. Platypus walk up the creek bed and find homes in our dams. But as we watch the waterways and forest with joy, we also know that these animals live and die as nature intended. They suffer with little or no human intervention. They play their part in the cycle of life, both predator and prey. And in many ways, they inform our views, because in *their* world there are no animal activists and no factory farmers. There's just the day-to-day scramble to survive. Eat, breed, flee or fight.

Many times, too many times, death comes without our intervention, even with the farm animals. I've spent hours helping the vet pull a too-long dead, half-rotten calf from the uterus of its mother. I've wept in the paddocks when a lamb has somehow lost the ability to suckle, and I've tried in vain to bottle-feed it, only to see it fade and die. I curse myself for not seeing the death coming, and ending the suffering earlier. We've had cows fall sick from snake bite that have had to be put down. We've lost pigs to crushing by their mothers, lost to drowning and hypothermia. We've lost lambs to exposure, chickens to quolls, cattle to illness. We've had piglets snatched from the paddock by wedge-tailed eagles and sent to a possibly slower death than we would give them. I have cradled the head of our first dairy cow, Maggie, as she succumbed to black mastitis over a period of days, before we finally realised the battle was lost, and called someone to end the pain and bury her body. We've fed turkeys and chooks with eye-droppers to try to nurse them back to health (most often unsuccessfully). Death like this hurts us. It nearly breaks our spirit. It stains our souls in a way that's hard to explain.

And yet where does most suffering happen on our farm? Where does most death occur? Well, apart from the garden, where most lives are lost (but most are insects, a topic for later in the book), it's in our wild parts. I see wallabies with disease or old age looking frail in the riparian areas of the farm, the fenced bits around the waterway and in the bush. I know that the magnificent local wedge-tailed eagles may swoop down and run the sickened animals until they drop. Or perhaps it will be the goshawks and crows who will try to peck out the eyes, leaving the stricken marsupial to die a slow death, ready for the carrion eaters' next meal. At night there's added worry for any wildlife from the local ground-dwelling carnivores. We know the quolls, opportunistic predators who usually eat carrion, may take out a bandicoot or a baby wallaby, or those too old or sick to run. Boobook owls patrol the space between the trees. Possums are omnivores who will rob birds' nests of eggs and young. We've started to see kookaburras, introduced to Tasmania a century ago, so not part of the original landscape – birds that despite their infectious laugh are prone to kill the young of other birds, mostly while the chicks are still in the nest.

But here, today, the death I'm about to describe is at our choosing. We've picked out a 24-month-old steer for our table, an animal of full size and with enough age to have great flavour.

I alter in my being as the gun is brought out. So does the slaughterman, with a straighter back, and higher shoulders. He becomes a different person; serious, intent. Or at least a different version of the same person. We've chatted about this on other days: his altered state when he kills, so he can remain a stable human being in all the hours when he's not behind the rifle and knife. He sets aside human norms so he can do the job at hand. Putting down an animal, slaughtering a living thing. Murdering an animal, in vegan speak. Killing another being so we can have food.

This is, at its heart, what we have chosen to do, and what some choose not to do. I may not pull the trigger, but I am as much complicit, if not more complicit, in this act as the man behind the gun.

I do feel guilty asking someone else to do this task. I feel that if I'm truly in touch with what I eat, this task should be mine. I think of myself as just another hypocrite because I'm always saying we should take responsibility for the food we eat. But killing an animal of this size – well, I worry that I might muck it up. That I might damage the steer without taking its life. I worry that I'll prolong life with a badly aimed shot and actually create suffering. So I bring in someone with more practice than me. More experience in removing life than I have had. Someone who won't mess it up.

Abrogating responsibility for this deed, having a surrogate do this act for me, makes me no better than my city self. At heart, I'm possibly no different now, really, than when I lived in Melbourne and rode the tram from omnivorous meal to butcher. No more personally responsible for the death of a conscious, warm-blooded being than I was when I worked for *Vogue* and *Gourmet Traveller* in Sydney, and published recipes promoting the eating of animals with never a thought to the farming system that they lived in. And yet … And yet.

I *am* more responsible. I *do* know what is happening in my name. I have made the call, readied the paddock, prepared the coolroom and compost heap. For this one animal and the end of its years, I am the one who has decided to play god. And this part, knowing of the life, the date of death, the use to which we will put this meat, I am completely cognisant and comfortable, at least intellectually. I have chosen the day this animal will cease to live, and while I will not personally wield the bullet or the blade, it only dies because I have chosen to eat meat. And I am not only okay with that. I'm satisfied within my skin that what I do is actually fine, and better than some of the animal suffering that goes on in all people's names, whether or not they choose to eat meat.

That said, when you see the soft brown eyes of the animal that must die, you do feel remorse. You feel shame. You feel dirty. There's no denying that the human ability that allows us to show empathy, to intellectualise, to feel emotions possibly beyond those of other animals – that ability *does* make killing another warm-blooded animal one step beyond. Where empathy starts and anthropomorphising – the placing of human traits on animals – begins, I cannot tell.

I never thought I'd be here. It was never my dream to have to help gut cows, to shovel the just-spilt blood from a newly killed steer into buckets. I don't think any abattoir worker in the country went through school thinking their dream job was at the end of a bolt gun with the smell of the kill floor drifting up from below. What happens at the end of a farm animal's life isn't something anybody except farmers, vets and slaughterers have to think of in the day-to-day. And no matter how many times it happens here, or happens offsite at our command, I don't feel great about it.

But what those who aren't at the kill need to understand is that animals die every day. They've died at the wheels of our car, on our driveway, when we've been driving at 15 kilometres per hour. They die when we hit a flock of birds that veer up from a crop in the country. They die when your neighbour has to poison the rats that threaten to eat through their electrical wires and burn their house down. Animals die all the time at the claws, beaks, incisors and talons of other animals. Death surrounds us. It's just that this one death is totally my responsibility.

—·—

The steer ambles into the paddock. It's chasing hay, the way a child chases a lolly. A steer, a castrated male, is a good beef animal – and Kevin, as we have named this one, has the perfect shape, despite his maternal dairy genes. Beef cattle are built like a brick. Squared, with flat backs and sturdy legs, they stand in contrast to dairy cows, which are all fat stomachs, big udders, protruding hips and visible ribs. Dairy cows are a fermenter on legs, particularly modern breeds that are required to give 40 litres or so of milk a day. Beef cattle are bred solely for meat, hence the boxy shape, with more muscle covering bone. Kevin, who is half meat breed, half dairy, has inherited his dad's physique.

Kevin leans down to take a chomp of hay, and a high-calibre bullet takes him down. There's a spot between the ears and eyes where a tiny piece of lead can completely stun the animal, leaving it to feel, theoretically, no pain. Kevin falls immediately. The slaughterman presses his finger to the eyeball, a low-tech but scientifically valid way to tell if the beast has actually been stunned, if the bullet has done its work: when you touch the cornea, and there is a blink, it means the animal is still conscious. Kevin's throat is then slit to bleed out the meat, as the heart keeps pumping post-stunning. You always let the blood run out of animals after being stunned, otherwise the meat can be unappetising or even ruined, be they wild venison, wallabies, chickens or cows. Kevin's legs move a touch. His back arches slightly. These movements, to the untrained eye a sign of life, are remnant electrical impulses; they don't signify that Kevin feels pain, or is still conscious, but rather that there are connections between the central nervous system and the muscles they control. These twitches can continue for minutes, but in this case, Kevin soon appears static. They unnerve me, though, these movements of an animal that is bleeding into the earth. They unsettle me, and I'm already a little stretched emotionally.

On our farm we used to say that our aim in rearing livestock was to give our animals a good life with only one bad day, meaning that they lived a life with as much joy and stimulation and ability to express instincts as possible, but the bad day was when they died.

Our aim now is that the animal only has good days (an impossibility, as it is with a child, but a noble aim it is, we believe), and the bad day is bad for us, but the animal doesn't know they're going to die. This is only possible on the farm. If they have to be trucked, we know our livestock don't like the journey. We also know they don't like the abattoir – especially ruminants, whose only knowledge of the ground has been grass beneath their hooves, and the smell of the farm and the bush.

At the abattoir, be it local or five hours away (and we've used a few), the animal stands on concrete or grated floors. They smell other, unfamiliar livestock, which plays with their sense of self and security. They probably smell fear, and they definitely smell death. I can see it in the reaction of our sheep and cattle at the abattoir. Their ears drop, their breath quickens and deepens. They move their lowered heads in strange ways and they poo in a manner they can't control, a sure sign of stress. The only farm animal for whom the smell of blood at an abattoir probably isn't so confronting is pigs. They're an omnivore, and the smell of blood could possibly mean dinner, not death, to them. But from our experience of dropping pigs off at abattoirs, they know what is coming, and they don't really want to be there, despite the vague promise of new food.

After a matter of minutes, Kevin's body has been bled out. Steam lifts from his hide in the cold morning air. He's hoisted up, feet first, by the crane on the side of a ute, then gutted onto the grass. The amount of guts is enormous, and the half-digested grass that fills his rumen is emptied into the paddock, the intestines and stomachs (they have four) loaded into a tub. This rumen (the first stomach) can be over a hundred kilos in a full-grown animal, filled with its grass and water brew. Despite the cold, there's the metallic smell of blood in the air. I can smell cow hide, and cow poo, and blood. As the guts are spilled the smell worsens. There's high, almost ammonia, notes, and a bitter tang. Gone is the scent of a living beast, which I find quite glorious and comforting, and in its place something rank and disconcerting. There's bile, of course, and all the smells of digestion and metabolism.

Kevin is decapitated then skinned, the leathery hide stripped from his body in a series of well-made cuts. This skin will be salted and exported to a tannery in some far-flung country, ready to be made into boots or saddles or belts, and sold back to us or others in well-off nations.

By now, it's not Kevin, it's a carcass. Through his slightly yellow fat (being half dairy cow means he was better at capturing beta-carotene from grass), the muscles still twitch in patches, something you couldn't see through the hide. A reminder of a life recently lost. The body is quartered, hung on hooks, and moved into the coolroom, where it will sit for three weeks before being butchered.

In the space of an hour, a 450 kilogram steer has become 220 kilograms of carcass. We've moved from animal to beef. From a living beast to meat.

For the next hour I'll be moving the guts and using them and copious amounts of sawdust and old hay to make compost. We'll cycle the nutrients back to our soil over the coming months. It's discomforting moving the head, so large, so lifelike, so heavy. It's in this moment, dealing with the entrails and the head, that I feel most shame at what I've been part of. How could I end an animal's life at its prime? When Kevin could, in theory, have lived another decade at least? Why now, why today? Why am I allowed to make this call? I can't answer that today. I made this decision, and other decisions like it on numerous days. I've chosen to eat meat, and in theory should have no guilt. But it's a grim day any day an animal dies on our farm, and even grimmer for me if it's in my name.

And it's in this time, on the day of the kill, that I feel most responsible. Not just for the death, but also (and importantly) for honouring the life. I keep the heart, the liver, the cheeks and tongue, and promise to cook them, though sometimes the dog does better than me out of what Italians would call the 'fifth quarter'. If I fail to use all the offal, in part because I'm the only one in our house who will eat it, I certainly have no intention of wasting the meat. I brine and slow cook the tongue. We might turn the heart into tacos, and slow braise the cheeks with a bottle of red wine, star anise and orange peel, reminiscent of a French daube. If you've ever been at a kill, and have any warmth to your soul, you'd have to lack a certain something if you didn't want to use every last bit of the beast. Having seen this animal from the day of his birth, having watched his grazing, his health, his herd, I know what is involved in producing meat. I have also seen the ultimate price paid by Kevin so we can put meat on our table.

Every day, millions of animals die. Some die like Kevin did, so we can eat meat. Many others die so we can live in the way we do, clothe ourselves and feed ourselves grain and pulses. They say that if we had abattoirs with glass walls, a whole lot more of us would be vegetarian, and I dare say that's true. Being witness to a kill can change a person, and possibly for the good. It changed me – but it didn't put me off eating meat. It does mean, though, I'm now far more cognisant of where my meat comes from, and how it died.

The reason I still eat meat, however – despite the cocktail of shame, guilt, and the conflicted emotions that I feel at what happens when an animal dies – is because I've seen how farming works. And how farming works may surprise those who have never really considered its wider impact.

CHAPTER 11

Think Being Vegan Hurts No Animal? Think Again.

There's a lot to be said for veganism. For the thinking eater, being vegan gets around a whole bunch of ethical grey areas. If you care about what you put in your mouth, the simple vegan argument is probably the most black and white way to choose how to approach the whole meat thing. There are no grey areas about so-called 'ethical' meat, or questions over exactly how 'free range' are the hens when there are 10,000 chickens to the hectare. Not eating meat, not buying products that come from animals – surely that means you're doing better not only for those animals directly affected, but also the environment, and your health as well? It's win, win, win, really.

But while the movement to utilising no animal products is on the rise in Western nations, it's still far from mainstream. Why, then, is it so hard to convince people of its worth if it really is a win all round?

The vegan philosophy is, at its heart, quite often about reducing suffering. By not eating animals, you – by definition – reduce suffering. It's a lovely idea. And I wish it were that simple.

Let's start with peas ...

Collydean (not its real name, but a real farm) is a 2700 hectare mixed farm in northern Tasmania. They grow beef cattle, some sheep, do agroforestry, have barley and some years grow peas. A LOT of peas: about 400 tonnes a season. And to protect the peas, which end up as a frozen product, they have some wildlife fences, but also have to shoot a LOT of animals. When I was there, they had a licence to kill about 150 deer. They routinely kill about 800–1000 possums and 500 wallabies every year, along with a few ducks. (As a side note, and in my mind to its credit, Collydean only invites hunters onto its farm who will use the animals they kill – for human food, or for pet food – and not leave them in the paddock, as most animals killed for crop protection are. They hate seeing the animals go to waste.)

So, over 1500 animals die each year to grow about 75 hectares of peas for our freezers. That's not 1500 rodents, which also die, and which some may see as collateral damage in the big picture. That's mostly warm-blooded animals of the cute kind, with a few birds thrown in.

Collydean's owners assure me that it wouldn't be financially viable for them to grow peas without killing animals. Which means that every time we eat peas, farmers have controlled the 'pest' species on our behalf, and animals have died in our name.

So it's not just meat eaters who are responsible for the deaths of animals. And as far as crops go, peas aren't the only culprit. I remember filming for *Gourmet Farmer*, a telly show on farming in southern Tasmania, when a heritage orchardist quickly moved an occupied possum trap out of view. Apparently the only way to protect his orchard from the devastating nightly possum attacks was to trap them. And the dead animals were then fed to the dog. But over 100 possums caught a year, from one small orchard, and pretty soon the dog had to go on a diet. This number of possums destroying an orchard isn't unusual. Another orchardist I know killed 120 possums one spring, an average of one a day, to protect his crop. If we eat apples, or drink apple juice or cider, then we have blood on our hands.

Looking further afield, the number of animals that die to produce vegan food is astonishing. Consider wheat, a common crop in Australia, and you'll also find a bunch of animals dying. And this time, let's look at the nutrient density of the food in question, because not all foods are created equal. According to an article by Mike Archer, Professor in the Faculty of Science at the University of New South Wales, roughly 25 times more sentient beings die to produce a kilo of protein from wheat than a kilo of protein from beef. Thanks to monocultures, and mice plagues, and our modern farming systems, a hell of a lot of small animals die to produce wheat. Yes, most of them are rodents, but a life is a life – and surely in the vegan world, all warm-blooded life should be honoured equally?

A plague is defined as a time when there are over 1000 mice per hectare of wheat. With Western Australia boasting 4–5 million hectares of wheat, that's potentially 5 billion mice in Western Australia alone, of which, according to Professor Archer, 80 per cent die when poisoned. And each wheat-growing region of Australia has a plague, on average, every four years. Using these numbers as a guide, on average 1 billion mice are poisoned every year just in Western Australia (averaging plague and non-plague years). Do we need to kill them? According to a 2005 Senate report, if we didn't kill mice, the cost of food would rise drastically; even with heavy baiting programs, mice alone cause about a $36 million loss from the Australian economy, so it's not just wheat where there's a problem.

Let's look at birds. In 2014, it was estimated that 500,000 geese of various species would be killed in Holland to protect crops. Not to protect other meat species, but crops. These crops, many of them, end up on Australian supermarket shelves (check your frozen spinach, or just about anything not fresh from ALDI).

Over a five-year period up to 2013, rice farmers in New South Wales killed nearly 200,000 native ducks to protect their fields. That's right, to grow rice. That's in addition to the animals indirectly affected, such as those that once thrived in the waterways drained by such a heavily irrigated crop on a dry continent. Ever been to Thailand and seen the roadside stalls selling frogs from the paddy fields? Or those in Vietnam where they sell frog and paddy-field rats, cooked over open coals? If you eat rice, any rice, you have blood on your hands.

That's how farming works. To grow something, other things are affected. Sometimes it's an animal, sometimes it's a helluva lot of animals. The most animals that die on our farm are the snails and slugs that we've battled to control, but would destroy our garden if left unchecked. We kill close to 5000 moths, slugs and snails each year to grow vegetables, and thousands and thousands of aphids. And we also kill 40–60 pigs a year.

Insects bear the brunt of all annual vegetable production. And the most exploited insect of all is the European honeybee. True vegans don't eat honey because it's the result of the domestication, and utilisation, of the European honeybee. They don't eat it because eating honey is 'stealing' honey from the hive, and because bees die in the process of beekeepers managing the hives and extracting the honey. And they're right, bees do die in that process.

Problem is, honeybees are very, very good pollinators, and a whole heap of crops are pretty much reliant on these bees to produce fruit – and even more crops would suffer from far lower production due to poor fertility if we didn't have bees. About one-third of all crops, globally, benefit from direct interaction with pollinators, of which European honeybees are by far the most efficient – so whether we eat honey or not, we are the beneficiaries of the work of the domesticated European honeybee.

Honeybees pollinate apples and carrots, melons, cucumbers, brazil nuts, pumpkins and zucchini, to name a very few examples. In their absence, some crops would come close to failure, and others increase substantially in cost. Gobs of bees die every year doing the work of pollination for us. And by gobs, I mean hundreds of millions of them. According to *Scientific American*, up to 80 billion domestic honeybees are estimated to have a hand in the Californian almond industry each year, many of which die both during the management process, and during the long journeys to and from the large almond orchards. Estimates go as high as half the bees involved die. And that's the carnage from just one crop.

What about vegan wine, you say? Vegan wine doesn't use fish bladders, or milk extracts, or egg as a fining agent (ingredients used to clarify many wines, beers and ciders). But don't forget the harvest. Come with me to watch grapes being picked, watch as huge tubs of plump grapes are tipped into the crusher along with mice, spiders, lizards, snakes and frogs. Sadly, vegan wine is a furphy.

Let's move on to peanut butter, that wonderful practical protein staple. Do you know how many parts of an insect are in each jar?

According to *Scientific American*, we all eat about 0.5–1 kilogram of flies, maggots and other bugs a year, hidden in the chocolate we eat, the grains we consume, the peanut butter we spread on toast. According to US regulations, which are easier to access than Australian data, pasta may contain an average of 125 insect fragments or more per 125 grams of pasta (a single portion), and a cup of raisins can have a maximum of 33 fruit fly eggs. A kilogram of flour probably has 15 grams of animal product in it, from rodent excreta to weevils to cockroach legs.

I don't bring this up for the 'ick' factor, but simply to show the true impact and cost of food production. When you eat, you're never truly vegan. When humans grow and process food, any food, other things die – and often we eat them.

Vegans can make a choice not to *knowingly* consume animal products, but it does seem food production gets unfairly singled out for killing animals, when every human activity has an effect on other living things.

Every hour, on average, on every day of the year, 32 native animals are killed on Tasmanian roads (and that's the lower end of the estimates, some say nearly double that). That's just one Australian state, with a population of about half a million. That's a lot of death, just so we can drive around in a tonne of metal from home to the shops, to work, or on holidays. Half of those wallabies and some possums will have joeys in the pouch, because wallabies always have a joey in the pouch. So it's probably more like 40 animals an hour, or over 350,000 a year, that die on Tasmania's roads. Some estimates, including those endorsed by the Royal Automobile Club of Tasmania, say 500,000 animals a year are killed. Should we get rid of our cars?

What about the 1200 brumbies – wild horses – due to be culled in the Snowy Mountains as I write? If they're not killed, there'll be less food and a more impacted environment for native animals, which will be more likely to die because the brumbies are left to live. We have to make a choice: one big animal, an invasive animal, versus countless small marsupials.

What about the hundreds of wallabies that die on Maria Island, a wonderful national park off the east coast of Tasmania, when wildlife management officers do a cull? The cull was first instituted in 1994 when tourists began complaining about seeing emaciated, starving wallabies on the island. Nature at work is brutal, but it wasn't a real drawcard for the island, watching survival of the fittest in real time. So behind the scenes, 3300 native animals were culled in that first year. Messing with nature is messy. We can't bear to see wildlife starve, so we close the island and shoot them when the public can't see. Culls have dropped to 500 or so animals in more recent times (and some years to none), as populations of wombats, possums, forester kangaroos and wallabies have come back more into balance. But culls continue because we want nature to be clean and pure and not cruel. But nature *is* cruel.

It's important to keep this in mind. These numbers – roadkill, the Maria Island cull and the Snowy Mountains brumby cull – have nothing to do with food production for humans. But animals still die.

Did you know that between 2009 and 2017, American wildlife officials killed 70,000 birds so that planes could fly into New York after birds caused a jet to crash-land on the frozen Hudson River? And that at least 13,000 birds are killed in mid-air when they're hit by US planes each year. Or that in 2013, Schiphol airport near Amsterdam killed 10,000 birds, mostly geese, to ensure planes could land and take off safely? Here in Australia, 100 ibises were killed at Perth airport in early 2018 to prevent bird strikes on planes.

So we kill animals when we drive. We kill animals when we fly, or transport goods by plane. We kill when we build railway tracks, when we farm grain, grow apples and mine sand. We alter ecosystems when we put up new housing developments, build bicycle factories and ship lentils. We push native animals out of their environments all the time, with the resultant pain and suffering you'd expect.

We set up simplistic ecosystems when we grow stuff, creating mouse plagues and swarms of locusts, then commit wholesale slaughter on the pests we have allowed to thrive. This is us, another animal in the complex web of life. This is the way humans are, just like all other species, where every action we take has an effect we may or may not appreciate.

Killing an animal for food or fibre is a small effect. Bigger is the ecological footprint of livestock on the land. But bigger still, and more destructive still, is the growing of plants for food, thanks to topsoil loss, the legions of animals killed to maintain monocultures, and the use of artificial fertilisers and the armoury of chemicals available to the modern farmer.

How you feel about this impact, these deaths, depends on your point of view. What doesn't change with your point of view is the fact that they still die. As American scientist Neil deGrasse Tyson, host of the TV series *Cosmos* says, 'The good thing about science is that it's true, whether or not you believe in it.'

Some may argue that all this roadkill, all that culling of wild horses, all the things that die in our efforts to house and clothe ourselves – all these creatures are going to die anyway: so why would you *increase* that number by choosing to eat meat? I think we can do better in so many ways, and perhaps being vegan is one of them for some. But, really, meat eating makes you no more culpable than those who fly.

A vegan world, in the short term, just isn't going to happen. Not only are people eating more meat in developed nations, they're eating way more meat worldwide. And even in affluent nations – places that can afford to focus on animal welfare and the ethics of eating – it's still only about 1 in 50 of us who have given up meat. The rest of us seem to be eating more meat than ever.

Peter Singer, academic, author and philosopher, is arguably the foremost proponent of 'speciesism' – the idea that we discriminate against other species, favouring some species over others, and treating different animals differently, even within the same species. Singer is also the co-founder of Animals Australia, and has been a bestselling writer on veganism for decades. Yet even he admits it's better to get the animals that *are* raised to have a better life than to suddenly expect the world to give up its love affair with meat.

Being vegan might be ennobling for the individual, but often it leads to powerlessness at the pointy end of farming. Not partaking of the animals that 99 per cent of the rest of society eat means you have no financial clout. Demonising meat eating, and this is the sad part, alienates the very people who are most likely on the same wavelength about animal welfare, but still choose to eat meat.

Veganism might be on the rise, but it's a slow rise from a small base. Animal suffering is happening now. Climate change, which some animal farming can influence, is happening now. Waiting for the world to turn vegan, when the global population is eating more meat than ever, isn't driving the change that is needed.

I've thought a lot about the issue, and I feel no guilt being a meat, fruit and vegetable farmer. What matters isn't that you or I eat meat, but rather how those animals live, and just as importantly, die. What matters is how we raise animals, how they spend their time, and also what we do with them once they're meat, and not living beings. It's about respect for the environment, for the farmer, and for the farmed. And that probably means eating less meat, in affluent nations at least.

I have been fortunate enough to be on all sides of this debate. I've experimented with vegetarianism. I've thought about becoming vegan. I've been to intensive chicken and pig farms. I've 'smelt money' and seen despair. I've also raised animals, killed animals (both wild and domesticated), and cooked animals.

And what I've found is that the animal world isn't isolated from the world of plants, and the place for nuanced, sensible debate about meat consumption should sit firmly with all, including with the omnivores of this world – a debate in which condemnation, aggression and intolerance should play no part.

Perhaps, for those not interested in eating meat, or who choose not to eat meat, it's about context. All the creatures killed in the raising of crops – the rodents, the insects, the birds – these are just collateral damage. This line of thinking is based on the fact that meat eaters (or their agents – namely, the farmers, slaughtermen, butchers and chefs) 'choose' a victim, so this is different to an animal dying as a result of random chance. But a death is a death. Suffering is suffering, regardless of whether a human was involved, directly, or not. *All* impacts of our actions need to be considered. And this I think goes to the heart of the matter.

What actions produce the least suffering? Some commentators believe that annual crops produce more suffering for more animals. The view is that life is life, that life begets life, and to live we must consume something that has lived, with impacts on other forms of life well beyond our circle of thinking. You eat a plant, and that affects an animal – one that was going to eat that plant (say a nut from a tree in the wild), one that dies because it was going to eat that plant (perhaps grasshoppers or caterpillars on farm crops), or one that might've lived in the wild if we didn't farm that plant at all.

All of us, vegans and omnivores alike, are the beneficiaries of the fertiliser and compost that come from either of two sources: animal waste or fossil fuels. Organic farmers use compost made from animal by-products, whereas conventional farmers use nitrogen fertilisers, which are produced using large amounts of fossil fuels. About 2–3 per cent of the fossil fuels burned each year is for making nitrogen fertilisers – accounting for roughly 3 per cent of the world's carbon emissions, including emissions from nitrogen released to the atmosphere. And then there's the global transport system, which uses fossil fuels to ship your Brazilian soy beans and Californian almonds all across the world.

If you don't use fertilisers made from fossil fuels, you need animal by-products. There's barely an organic fruit and vegetable farmer out there who doesn't use some kind of animal by-product (manure, blood and bone) or the compost that contains it.

And there's barely a farm that isn't reliant on gas and oil to make the fertiliser, run the tractors and ship the goods. Most estimates put it that the amount of fossil fuel needed to grow a calorie of food and get it to the table is ten times more than the food calorie itself. It's a negative-sum game. Grains and monoculture crops are worst amongst them – whereas grass-reared animals, killed and sold locally, are amongst the more efficient producers of food energy for fossil fuel use.

Take away the use of animal waste in the farming system, and things will swing further to one side. If you want truly vegan agriculture, you're going to have more fossil fuel emissions (see the next chapter), and in the process end up with more expensive food, poorer pollination and reduced variety thanks to the removal of domesticated bees.

There are other questions. Would it be okay to eat chickens that are bred to have no pain or stress receptors in their brain, if that were possible? Those chickens would never suffer, no matter how intensive the farming system.

What about the hormone replacement therapy compounds that come from horses (and the ponies) the mother mares have to keep giving birth to, in order to keep their hormone levels up?

What about the innumerable advances in science that come from experimentation on animals, including the horseshoe crabs that are milked for their blood, which is used to verify the safety of just about every injectable pharmaceutical? (The blue blood, harvested from half a million crabs a year, has a miraculous ability as the most sensitive indicator of bacterial endotoxins ever discovered, meaning that its use has made injectable pharmaceuticals phenomenally safe.)

Should a co-op that rescues layer chickens from intensive farms be able to sell the eggs to help pay for the upkeep of the newly freed chickens – and can vegans eat those eggs and remain truly vegan, given that these are eggs which will be produced regardless of whether people eat them or not?

And as we saw earlier, what are vegans doing about feral cats, which besides predating upon a staggering number of native animals every night of the year, also spread toxoplasmosis to other animals that then suffer in their wake?

Vegans are welcome to voice their opinion that raising and eating meat has consequences. Indeed, some of those consequences, from the personal to the animal to the environment, are worth some serious thinking about. It's quite possible that eating less meat might mean less suffering. But don't be fooled into thinking being vegan hurts no animal.

I'd like more people to join a sensible discussion about animals, their use for food, and the impact on them from farming and other human activities. What I want is a spirited, intelligent, respectful dialogue that includes meat eaters as well as non meat eaters and farmers. Sadly, up until recently, with extremists on both sides dominating the debate, all we've seen are worse outcomes for intensively farmed animals.

That conversation needs to encompass all our impacts, from crops sprayed with pesticides and herbicides, to our animal farming systems. Imagine all the death if you kill everything in a field that isn't cotton? Plants, animals, predators, birds and insects are all affected by weed-killers, as collateral damage. How much damage is enough? How much can we sustain? What is right, what is wrong? I don't really think I want animal rights groups like PETA deciding these issues for the entire community – nor the owner of a feedlot who relies on the cheapest grain possible.

What we need is less discussion of arbitrary personal choices, and more talk about animal welfare, about lives won and lost, about a definition for suffering. If activists want to help animals, there must be a better way. Later in this book I'll give my views on the best way to help, and it certainly isn't about hate, fear, blame and division. It's about hope, empathy, wisdom and care – both for the farmer, and the farmed.

The way forward is filled with light, not dark, with love not hate. It involves compromises and the recognition that your own personal moral compass isn't the only one in the world. The way forward is filled with hope.

But first, we have to look at what the problems are that we need to overcome.

CHAPTER 12

Beef Is Killing the World. What About Asparagus? Or Golf?

Have you seen *Cowspiracy* – the documentary that exposes how livestock, in particular ruminants and especially cattle, are bringing on environmental Armageddon (albeit with some questionable statistics)? Between it, and other shows, websites, headlines and commentary, you'd probably be aware that it takes 16,000 litres of water to produce a kilo of beef. And that greenhouse gas emissions from livestock outstrip those from all transport combined – including cars, trucks and airplanes. And that in 2006 the United Nations released a report titled *Livestock's Long Shadow*, which shows the cost of animal production, on a global scale, is simply too high.

There's no denying the huge environmental cost of beef, but before we delve into those, let's look more broadly at the environmental effects of different diets.

According to a 2014 study by the University of Oxford, meat-rich diets (which the researchers defined as anything over 100 grams per day) ended up with 7.2 kilograms of carbon dioxide–equivalent emissions per day. Vegetarian and fish-eating diets resulted in about 3.8 kilograms of carbon dioxide or equivalent per day, while vegan diets produced only 2.9 kilograms. The Oxford study looks at lifecycle carbon emitted, from growing to the retail stage, and found moderate and average meat eaters have about double the carbon emissions of vegans. (Their figures are based on conventional agriculture, not best practice farms, and don't take into account what happens at the consumer stage – where boiling dried chickpeas, for example, might produce more carbon than searing a steak.)

Food, generally, is responsible for about 20–25 per cent of all global carbon emissions, so even if we all became vegan, human impacts on the world through greenhouse gas emissions would only drop by about 10 per cent. We still have to eat, and through our everyday lifestyles we will still continue to emit carbon.

So while meat eating does, according to most research, produce higher emissions, vegan food has other costs that aren't apparent on simplistic figures like the above.

———

A farm, by definition, is a foreign landscape that we impose on mother nature. The kind of farming most Australians would recognise has historically – and sadly, inevitably – turned arable land into deserts, and sent species extinct.

In his book *Dirt: The Erosion of Civilizations*, American writer David R. Montgomery makes the argument that one of the limiting factors in building long-lasting human settlements of great note has been our degradation of soil, because farming increases the rate of topsoil loss.

Topsoil – that magical few centimetres (and more like millimetres in parts of Australia) – is the living, nutrient-dense bit that does most of the growing of plants. And at current rates, thanks mostly to wind and water erosion, according to Montgomery, the world is losing about 2–3 tonnes of topsoil per person, per year. That's a lot of topsoil being blown away or going down the creek. We're running out of arable land which we can ruin, with almost all of the world's usable land already under cultivation. There's no plan B when it comes to where to grow the world's food.

Now, let's be clear, topsoil is always lost. But few things erode topsoil more effectively than humans have. By cutting down trees to bare earth we speed up erosion hundreds if not thousands of times. And where we replace those trees with annual crops, which inevitably requires digging and exposing bare earth, we lose topsoil even faster. Overtrampling by livestock can also ruin or lose topsoil. Topsoil *can* be made, typically through the breaking down of deep rock into minerals, dying roots that break down into the soil, the action of microbes in the subsoil, and by the addition of leaf matter for carbon (ideally with a dose of animal manure for nitrogen on top). But this is usually incredibly slow, and we lose substantially more topsoil than the earth creates, thanks in most part to the widespread planting of annual crops.

We can only grow vegetables and grains on the scale we do now because we use very simple fertilisers made from oil and gas reserves. In other words, the petrochemical industry subsidises your food. Nitrogen for fertiliser is extracted from the air using a really condensed energy source, usually natural gas. (Our atmosphere is about 80 per cent nitrogen, but extracting this nitrogen takes a lot of energy, and about 3 per cent of greenhouse gas emissions come directly from this source.) But this simple nitrogen that is captured from the air, along with the phosphorus and potassium (NPK) that are added to simple fertilisers to make plants grow much bigger much quicker, aren't very well bound up in soil. Most of these nutrients wash out of the soil through the normal action of rain – and far more easily than they do from natural fertilisers such as compost. And to complicate matters even more, the nitrogen that is washed out from agricultural fertilisers becomes a potential toxin, leading to increased nutrient runoff that can poison land and waterways downstream.

So growing annual crops – particularly, but not exclusively in monocultures – which exposes soil and relies on petrochemical fertilisers, is a pretty bad way to look after soil. Less damaging to soil are what are called perennial crops – those that live on from year to year, like tree nuts, and fruit, or perennial grasses for grazing. In perennial agriculture, if managed correctly, there is far less soil disturbance, and less erosion, and less topsoil loss. There's even the potential for topsoil gain.

Good multi-species farming, the sort humans mastered a few centuries ago, manages rotations of grass and/or grazing (for animal manure, nitrogen fixing and carbon sequestration), with annual crops and periods of fallow (rest). This cycle of rotated land use reduces nutrient depletion in several ways. It means nitrogen can be sucked from the air and stored in the soil by bacteria that have a relationship with leguminous crops such as peas and clovers. Carbon is put back into the soil as the roots of grazed grasses die in the ground. The rich fertiliser from animal manure is incorporated back into the soil through the action of worms, dung beetles and microbes, and trampled grasses can compost into the top layer of the earth.

Topsoil loss, and its potential replenishment, are arguments in favour of perennial agriculture. They also work as arguments in favour of grazing with livestock. Well-managed grazing can return and store more carbon in the soil, improve soil health, and reduce runoff that can be associated with annual crops.

What's really nice about this, to the omnivore, is that our domesticated grazing animals are pretty good to eat, too.

———·———

Beef, in popular culture, has become synonymous with environmental vandalism thanks, in part, to cattle's methane emissions and water use. You can find vegan Pinterest boards claiming that if you go a year without using paper you'll save 8.5 trees, and if you go a year without eating beef, you'll magically save 3432 trees. I don't know how they came up with those figures (I'm guessing based on American grain-fed feedlot beef, and some loose maths), but on our farm we can eat beef every day of the year and still grow more trees.

The argument in the more believable mainstream science press is that livestock – especially ruminants (multi-stomached animals), because of their unique digestion – are killing the world. And in a way this is true. Except the animal most culpable isn't the cow or other ruminants such as sheep or goats. It's humans that are responsible. It's how we manage things that matters. Beef can have a higher environmental cost than some other meats, and certainly some vegetables, but not all cattle-farming systems are created equal. And removing ruminants such as cattle from the diet of everyone in the world is going to cause more problems than it solves, certainly in the short term. (I cover this more in detail in Chapter 16, How Do We Feed the World?)

Let's take a look at some of the problems associated with beef production. Methane emissions are one good starting point. Methane is a particularly potent greenhouse gas, with an ability to store the earth's heat that is many times stronger than carbon dioxide. When a cow or other ruminant eats, their bodies can only digest the tough cellulose and some other elements in grass because of the fermentation action of microbes. There are zillions of bugs in a ruminant's gut – about 50 billion bacteria, along with protozoa and fungi, per millilitre. That's a quadrillion microbes per cow – a thousand million million microbes, just in the largest of a cow's stomachs, the rumen.

These microbes ferment and break down ingested grass, other leaf matter and seed heads – things we can't digest – into products a cow can digest. (A single-stomached animal like a horse or kangaroo only uses 70 per cent of the available nutrition in feed when compared to cattle, meaning they need 30 per cent more grass or fodder per kilo of weight gain.) The cow also benefits from the nutrients released by the microbes, and can access a lot of their nutrient needs from the microbes themselves, which can contribute 70–90 per cent of the protein in a cow's diet when they die. That's right, up to 90 per cent of a cow's protein needs are met by dead microbes. Who knew?

Anyway, back to those fermenting bacteria, some of which produce methane as a by-product. Cattle expel the gasses of digestion – methane and carbon dioxide included – mostly out of their mouths, and a fraction out as farts. Now this digestive ability is common to all ruminants, but cattle outstrip the others in their impact because of their huge global numbers.

Greenhouse gases, of which methane and carbon dioxide are two big culprits, are well-established disruptors in the atmosphere, changing the global temperature, leading to warming, extreme weather events and ocean acidification. But both gases have very different impacts over time.

Some of the best estimates put the warming potential of methane, on its initial release, at 67 times worse than carbon dioxide (some say 84 times worse) – but methane is also very short acting, being removed from the atmosphere in 12 years (versus carbon dioxide's potential to linger for thousands of years if not stored by plants or the ocean), and reducing to just four times more damaging than carbon dioxide after a hundred years.

In other words, methane is an incredibly potent greenhouse gas – but not in the very long term. Removing methane from the atmosphere may provide a quick fix to our climate change woes, but only while we work out how to mitigate the damage from all the new carbon being added through the burning of fossil fuels.

Carbon dioxide, released when forests are cleared, grasslands are badly managed, and most importantly, when we burn fossil fuels such as coal, oil and gas, has far more damaging potential not just over the next few years, but for millennia to come.

Now, methane is an issue, and we probably should reduce the number of cattle around the world, especially grain-fed cattle in feedlots, but the contribution of methane to the global warming potential of the earth is only about 5 per cent, compared to other greenhouse gases. And it's important to remember that methane – which is one carbon molecule attached to four hydrogen molecules, CH_4 – is only constructed from carbon already in play in the world. If we weren't burning fossil fuels, carbon would be cycling fairly evenly between the atmosphere (as methane, but mostly as carbon dioxide), in plants, the oceans and soil, and stored in the bodies of living things like animals.

As the Food Climate Research Network puts it in their very well-reasoned study, *Grazed and Confused*:

> *It is important to underline the point that livestock*
> *add neither new carbon nor nitrogen into the system.*
> *They merely contribute to their accumulation in*
> *some compartments (reservoirs) along the cycle:*
> *in soils, or in plant and animal biomass.*

In other words, what we *should* be worried about is new carbon being released into the atmosphere – carbon that has been safely locked up for millions of years.

Because it persists in the environment for so long – about 40 per cent will linger for thousands of years – any new carbon dioxide released into the environment will have a far stronger cumulative effect on climate change than methane. Any new methane (assuming animal production doesn't increase, but the gas emitted today is added to the gas emitted yesterday) means the greenhouse potential *persists*, but doesn't *increase* as drastically.

It's technical, but important, because while reducing methane from livestock would be a really good thing (and remember, there's also methane from wetlands, from fracking, from coal mining, from natural gas extraction, from dung – particularly intensive-farm dung – as well as from rotting food), the only way to counteract long-term permanent change in the climate is to address both methane and carbon dioxide from long-stored carbon.

Reducing the number of cattle on earth wouldn't hurt, but looking at what fossil fuels we burn to rear them wouldn't be a bad thing either.

The methane that grass-fed cattle and other ruminants exhale comes, paradoxically, because they can digest things that aren't edible to humans. So while we may say that producing beef is 'bad', we can also say that by eating grass, or hay, or silage, or any number of crop residues that humans can't digest, cattle are producing human-grade food in the form of meat or milk that we *can* digest. They produce human food from non-human food. All this for a net-zero impact on carbon in the system.

What we mustn't do, anywhere, is continue a cycle of harm: to burn fossil fuels, and denude forests to create grazing or cropland, hence releasing even more carbon, to feed even more ruminants.

The problem comes with increasing carbon in the cycle – namely the fossil fuels used to feed, house and transport those ruminants. Feeding *grain*, rather than grass, to livestock that can eat things humans can't is a pretty dumb way to feed the world. And if 80 per cent or so of the grain grown in the United States is fed to livestock, which it is, then that's a dumb way to run a food system.

But here in Australia? Or globally? What does that mean for the rest of the world?

Since 1996, when the Argentinian government approved genetically modified soybeans, the country has cut down a quarter of its native forest to grow soy. That's a lot of carbon released from a former forest. Most of those soybeans – a rough 80 per cent or so – are fed to livestock. Not just cattle, but chickens, pigs, and even farmed fish. The other 20 per cent, and some of the oil extracted from the 80 per cent, is used for humans. So this style of farming, of denuding forest to feed grain to livestock, adds a lot of carbon dioxide to the system.

It's not just the Amazon and far-flung places that have an issue with releasing carbon. Our very own state of Queensland hasn't really done much better. Sadly, Queensland was, until recently, famed for being the only region of a developed nation to cut down a football field of trees *every hour* in the interests of producing beef. A million hectares cleared over a single four-year period is a sorry tale. It would make clearing in the Amazon look paltry, except, sadly, the Amazon's clearing is worse because it's even bigger in scale, and its destruction releases way more carbon per hectare thanks to the much thicker vegetation that was removed.

In Queensland, it was a simple change in state government policy that led to such short-term thinking and such environmental vandalism. Unless trees are allowed to grow back, this is carbon that has been released to the atmosphere for millennia. The law changed again in 2018 to restrict land clearing, but another reversal could again see things slide. In the long term, if all these trees in Queensland are allowed to grow back, we will have stored the carbon again that was released. But in the process we have eroded more topsoil, which will be enormously hard to restore, and have also done long-term damage to the local environment – as well as the environment further downstream, including the silt that is affecting the Great Barrier Reef.

Of course, the beef industry will tell you what good environmentalists they are. On their website goodmeat.com.au, Meat & Livestock Australia, the peak beef and lamb industry lobby group, reckon: 'Australian livestock farmers care about the environment in which they raise their livestock.' Yet some of those very same farmers have chopped a hell of a lot of trees down, in marginal country, in catchments that drain straight into the Great Barrier Reef. Their livelihoods, at least in the short to medium term, don't depend on a thriving environment and healthy ecosystems. I've driven and flown over this country, including heavily cleared parts of central Queensland. I've studied the satellite images of deforestation and land clearing done in beef's name. And if those farmers think beef isn't killing the world, they need to take a good long look in their own back paddock.

Graziers, in a perhaps misguided effort to get more grass to grow in the poor soils and marginal country of outback Australia, think trees are standing in the way of more cattle. But every time they cut down trees, they reduce the fertility of their land. Cattle barons, and increasingly cattle conglomerates, are mining the nutrition of some of our most marginal country, and doing future farmers few favours.

Now, let's deal with the 16,000 litre argument – that it takes 16,000 litres of water to produce 1 kilo of beef.

Sometime in the last decade or so, every slack journalist or science reporter, every animal activist around the globe, has latched onto this figure, which was based on very loose maths. Fresh water is a finite resource – so 16,000 litres of water to produce a kilo of beef is an appalling conversion in anyone's language. Surely if it takes that much water, then beef *is* killing the world.

But what *is* that water?

If it's rainwater, then it's not like it has been drawn from another source and diverted to beef. It's helpful to look at our farm, and see that not all farms, and not all farming systems are equal.

Our farm, let's say 20 hectares of it, gets about 700 millimetres of rain a year. And 1 millimetre of rain falling on 1 square metre gives you 1 litre of water. If you multiply the amount of rain that falls per square metre by the total rainfall for the year, you get a pretty big number.

I've done the boring maths here: 1 hectare = 10,000 square metres. So 20 hectares is 200,000 square metres. At 700 millimetres of rain per year, that parcel of land receives 140,000,000 litres of water per annum. That's 140 million litres of water. A lot, right?

On those 20,000 hectares, we produce about four head of cattle, sometimes more, each year to use for food, but let's work with that low number. Each beast might contribute 170 kilograms of meat to our kitchen, a total of 680 kilograms of beef a year. Divide those 140 million litres of water by 680 kilograms ... and guess what? You come up with the staggering total of nearly 206,000 litres of water to get a kilo of beef. *Way more* than the 16,000 litres often quoted.

Now, we also get milk off the animals that roam our farm, and nearly half the land is used for wildlife, but it sounds like a lot of water. The thing is, this water will fall to earth whether or not we run cattle – or whether we even use it or not. It will cascade down the hills and grow grass, it will fill the dams and spill down the creek even if we had apple orchards or blueberries. You can't measure all this rain and say it is used for animal production. Because not all of it is. This way of looking at water capture and usage is just too simplistic.

So, you might be asking, what about irrigation? Aren't you taking water that could be used for other purposes? Well, we do use water for growing vegetables and fruit, but we don't irrigate our pasture, and nobody downstream is missing out on the water. We know this because there's only a dry creek bed, and a huge dam on the neighbour's property, in the 600 metres or so between our boundary and the river. Yes, our cattle do need to drink, and this water is taken from those same dams that overflow each winter. The cattle drink it up, and pee most of it back onto the same paddocks. But let's add that drinking water to the total of, what was it now, zero (because we haven't irrigated or removed any water up to this point)? At 50 litres a day that a cow might drink, for 700 or so days (about two years on average), that's 35,000 litres of water used per animal. At 170 kilograms of beef per animal, we're now talking about 205 litres of water per kilo of beef produced. There will be some water use for the parent cattle, and some for the abattoir, but on our farm I'd argue we'd be looking at about 300 litres of water per kilo of beef at the maximum.

So where, oh where, does that ridiculous 16,000 litre figure come from? It's from a professor at the University of Twente in the Netherlands, who was also the scientific director of the Water Footprint Network, Dr Arjen Hoekstra. The aim, originally, was just to document our water footprint, much as we do carbon footprints. An apple: 70 litres of water. A cup of coffee? Perhaps 140 litres, to grow and process the coffee beans, package and ship them, as well as the water used to create the label on the pack (but not the water used to produce the actual cup of coffee). A single almond grown in California? According to some US figures, 4 litres of water for a 1.2 gram nut, which translates to about 3300 litres of water per kilo. (A kilo of wheat can take between 800 and 4000 litres of water to produce – but you don't see frightening headlines quoting this!)

Hoekstra's research allowed for a three-year-old animal, fed 1300 kilograms of grain, and included massive amounts of water that grew the pasture and hay. Headline grabbing it is, but what does it actually mean? In terms of red meat research in Australia, what are the actual figures, if we take rainfall out of the equation? In 2010, *The International Journal of Life Cycle Assessment* had a paper on just that, reporting that the amount of water needed in our southern climes was somewhere between 18 and 540 litres per kilo of red meat, depending on how you measured the water inputs. A long, long way shy of the supposed 16,000 litres per kilo in the Dutch paper.

———·———

If you take cattle, which can range and eat grass, put them into a feedlot, where you have to bring in water and food, and the grain and hay they eat is irrigated, *actual* water use rises exponentially. If you also factor in the need to hose out trucks and clean feedlots, and deal with piles of excrement that in a paddock would be considered fertiliser, water use may well even climb closer to the Dutch figure. But in an extensive grass-fed grazing system, water isn't necessarily a limiting factor, or an environmental factor, to worry about, even in more arid areas.

What matters is how animals are used in farming. And that's why *humans* are the problem. As we'll see in Chapter 16, grazing animals are best suited to places not workable for crops, or grazed in between crops when land is rested, or eating things we can't eat to produce even more food – nutrient-dense food – than plants alone.

But you can't just graze and get great results. Overgrazing reduces the amount of carbon trapped in the soil, as well as damaging the soil itself. Badly managed farm animals contribute to runoff, loss of biodiversity, and increased greenhouse emissions. Well-managed animals, on the other hand, can sequester more carbon in the soil, as grasslands are quicker to grow and store carbon if managed correctly and not overstocked.

The problem, of course, is with the human management of the system. The problem, and the solution, lies with humanity.

The single best thing you can do to ensure carbon sequestration (storage) in the soil is to not cut down trees. The next best is to definitely not turn grassland into cropland to grow even more grains – and then, if you do grow grains, to not feed those grains to ruminants. Sure, cut down your beef consumption, but don't eat too many grains grown on land that should be left fallow in the Amazon, either.

A little bit of beef, from animals raised on grass that is close to where the beef is eaten, raised on land that isn't suitable for other crops? That won't kill the world. It may help feed it.

But beef that is grain fed, from areas being deforested or overgrazed? That has the potential to bring all good farming undone.

———

I guess I often wonder why beef often gets headline billing as The Big Threat. Perhaps it's because of the American obsession with the hamburger, and the American obsession with grain-fed meat. I think of beef, and all meat, as an indulgence – as Simon Fairlie so eloquently captures in the title of his book *Meat: A Benign Extravagance*. It's a privilege to be able to eat beef, and it should be held in high esteem. We definitely should have fewer head of cattle on earth, and be cleverer about how we use that meat. But you don't see other products get anywhere near as much of the hate press as beef.

If you just look at land use, sure, beef cattle aren't always the best thing to grow in some regions. But is asparagus? A really intensive grower can get between 5 and 20 tonnes per acre per year. That's a crop that is left idle for 10 months a year, and needs watering, weeding and fertiliser to stay healthy.

Strawberries? You can grow, conventionally, up to 21 tonnes per hectare a year, if you use a toxic gas that was used as a weapon in World War I (chloropicrin, in tear gas), or a powerful ozone-depleting chemical (methyl bromide) to fumigate the soil, plastic sheets to smother weeds (sheets that end up in landfill after only a year or two), and spray it routinely with chemicals that wouldn't pass organic certification. Oh, and then rest the beds for a couple of years in between to avoid disease and pests – usually returning the land to grass (and, ahem, grazing animals on it).

If you take this good, fertile soil that grows strawberries, and instead raised milking cows on it, you can potentially get 20 litres of milk per cow per day, at an average of 2.5 cows per hectare. Over 10 months, that's 15,000 litres (15 tonnes) of milk. What's more a staple, milk or strawberries? What has more high-quality protein?

If it was beef cattle, you could run – only on high-quality ground – a cow per hectare, and get about 85 kilograms of actual meat (not bone or cartilage) a year if you kept them until two years of age.

Now, I know these figures can seem a bit eye-glazing, but they do show direct comparisons are hard to make unless you consider all aspects of a crop and the soil it grows in. You can run numbers on all sorts of things and go round and round in circles. The reality is, beef isn't efficient in terms of land use, unless it's land that can't be used for other purposes, or needs to be rested for other crops. Beef is, however, protein and nutrient dense. Here's some comparisons:

Strawberries per 100 grams: 33 calories, 0.7% protein, 8% carbohydrate

Asparagus per 100 grams: 20 calories, 2.2% protein, 3.9% carbohydrate

Beef per 100 grams (15% fat): 250 calories, 26% protein, 0% carbohydrate

Milk per 100 grams (3% fat): 42 calories, 3.4% protein, 5% carbohydrate

In truth, beef isn't an essential food, despite its complete protein and high nutrient density. It's a luxury food. An indulgence. But so are strawberries, and so is asparagus. And because we grow all of these on our farm, I can tell you there's only one of the three that is low in all input costs and is also virtually labour free: beef.

Of course, we can condemn beef as much as we like, or defend it as much as we like. People in affluent countries like Australia eat it because they like it. They eat it because it is culturally appropriate and culturally lauded. They eat it because it fits in with their life, their cooking skills, their needs. It has an environmental cost, and that cost, by most measures, is higher than that of other meats – and far higher than many other crops. But in the right system, with the right management, fattened on all those things we can't eat, beef can be a part of a sustainable land management system and our diet.

And we shouldn't forget that globally about 200 million pastoralists, many of them subsistence farmers, are reliant on grazing animals not only for income, but also for food. Speaking at the 2017 World Food Prize, Dr Louise Fresco put it like this:

Meat consumption in moderate quantities is important not only for human health but also for the health of the planet. That's largely because livestock, including

grazing animals, poultry and pigs, are top digesters of biomass. In classical commodity supply systems, the optimal production of human edible proteins per hectare of land without depleting resources or biodiversity stands at a ratio of 88:12 of plant- to animal-based proteins. But in a food system in which all available plant biomass is unlocked by using the great digestive capacity of livestock, this ratio shifts to 55:45. This means that we need less land and fewer resources to produce nutritious food, leaving more land for nature and urban development.

Dr Fresco is a great fan of circular food systems – where crop residues, grass and forage are cycled through animals to create protein for humans.

A linear food system is the one we generally have in developed nations, where we grow something especially for it to be eaten by humans – or the animal to which it is fed is then eaten by humans. It's not just beef cattle in this equation, but also, and very importantly, pigs and chickens (known as monogastrics, because they only have one stomach). Both chooks and pigs are great recyclers, with the ability to turn what we don't, won't or can't eat into pork, eggs and chicken meat. The problem comes when we don't use them as recyclers, but rather just feed them grain and supplements in order to get ever-cheaper meat.

———·———

If you want to find the real demon in terms of land use, greenhouse emissions and the amount of food created per acre, beef isn't it – because at least it serves a purpose in our diet.

If you want to condemn something for being a wasteful, unnecessary, carbon-emitting use of land, then really you should blame golf. Often built on good-quality, relatively flat land close to cities – land that would be ideally suited to farming – golf courses are fertilised and mowed using fossil fuels, usually sprayed with toxic herbicides, and often gated or kept only for the benefit of the few. Golf contributes nothing to our food system, only removing valuable land from being farmed, in return for a lot of emissions, a lot of waste, and a lot of not much else. So blame golf.

Or blame the car, which not only produces enormous carbon emissions, but in our modern cities mean roads take up about half the land area – space that was, and could be again, farmland.

Blame lawns, which in Sweden take up about half the suburban land area; lawns which are cut using fossil-fuel mowers and sprayed with herbicides. Or turf in the United States, which includes lawns, sports fields and golf courses, and covers four times more land than the growing of corn. Or lawns in Australia, which *The Sydney Morning Herald* calls 'Australia's most irrigated crop', using more water than anything else we grow for food or fibre.

Blame horses, grazing on things cattle can digest much more efficiently.

Or blame palm oil, or coffee, or chocolate, or pinot noir, because these all use land, they all have an environmental burden that is greater than vegetables, and they are all completely unnecessary, and still humans choose to pursue them. In 2017, around the world, there were 7.5 million hectares under vines for wine – an area bigger than Ireland, bigger than the whole of Tasmania, and slightly smaller than the Czech Republic. In the same way nobody has to eat beef, nobody *has* to ingest wine. (I personally, and unapologetically, pursue coffee, chocolate and wine on regular occasions – though I'm rather unimpressed with the world's dependence on palm oil.)

As we've seen, nothing comes without a cost, and it's up to us humans to make the best decisions about our food choices, in order to protect the environment that sustains us. Those choices will depend on where you live, what culture you belong to, your finances, how good a cook you are, your palate, and your own moral compass.

Everything we do has an impact, and much as it might suit some to find a scapegoat, we can't really just blame beef for killing the world.

CHAPTER 13

Antibiotics:
How Farming Is
Ruining it for Everyone

There's a book I read recently, a dense, almost impenetrable tome, called *Big Farms Make Big Flu* by Rob Wallace, an evolutionary biologist who specialises in influenza. And within its pages, carefully analysed and annotated, are the origins of swine flu. And the origins of the most recent bird flus. It makes frightening reading, especially when you consider the numbers. Wallace's premise is that the more we intensively crowd animals, the more we compromise their immune systems and place them in harm's way, and the more disease we create. While his musings are often on influenza – against which antibiotics are of no use – he also describes disease more broadly. He warns that the more antibiotics we use, the more resistant other bugs will eventually (and irreversibly) become against those antibiotics, and the more likely it is that we'll have a catastrophic outbreak of disease. The worst-case scenario is a superbug that may cross species, not just into other livestock, but – terrifyingly – humans.

Admittedly, Wallace's is worst-case scenario stuff. But consider the simple numbers. About 80 per cent of all the world's antibiotics are used on livestock, arguably only needed in such vast, vast quantities because of the concentrated nature of intensive farming. A quarter of all the world's antibiotics are used on animals in China alone, which produces about *half* the world's pork. As a consequence of such widespread antibiotic use there's antibiotic resistance in chickens, in pigs, in the bugs that live in beef. Antibiotic-resistant bacteria are being isolated on the bodies and clothing of abattoir workers and farmers.

At every turn, there sits a terrible spectre: the inevitable rise of the superbug.

It's important to talk about why antibiotic overuse is so problematic. It will get technical, but I'm going to delve quite deeply into this because it's critical to the health of the whole farming industry, and human health generally. Knowing about this stuff – just like understanding how the animals we eat are actually raised – is important to help keep the system accountable.

———·———

The way it works, simply, is this. Bacteria, many of which can divide into replicas of themselves every 15 minutes or so, have a very short life, and because any beneficial traits can only be passed on to every new generation, things that breed quickly are able to adapt really quickly, far quicker than species that have a long time between generations. If we use an antibiotic – 'anti' meaning *against,* 'biotic' meaning *life* – the aim is to kill bugs and bacteria, focusing on those that are harmful to us, or to the animals in our care.

But if you give an antibiotic to someone, and it only kills 99.9 per cent of the germs, because 0.1 per cent of the bacteria have a resistance to it, the resistant bacteria can breed up very quickly in the absence of the other bacteria, and we eventually end up with a superbug that is entirely resistant to that antibiotic.

We know this is always going to happen – it's the *speed* at which it happens that is the worry, because it happens at a speed that outpaces our ability to keep devising new antibiotics. (Not that we've developed many new classes of antibiotics in the last couple of decades.) These resistant bacteria can transfer their genes, including antibiotic-resistant genes, directly to other bacteria, hence speeding up the resistance potential. Why does this matter? Because for many decades antibiotics have been brilliant as a last resort against infection and disease, saving countless human lives. If antibiotics lose their effectiveness, many, many human lives are put at risk.

Yet despite these dangers, antibiotic use in animals is rife. In Australia you can feed sub-therapeutic doses of tylosin to pigs on a vet's advice – and I don't reckon there's many pig farms where the vet employed by the farm says no. Tylosin is fed to the piglets for about five days prior to weaning in most systems, and for about 5–6 weeks post-weaning in many sheds. Not because the pigs are sick, but ostensibly because when you wean a piglet at about three weeks old, they're not really ready for it, and a few die. Tylosin helps to keep those few piglets alive, so it is used as a prophylactic – as a medicine that is used to prevent a disease, rather than as a cure. So, being a prophylactic, every piglet in those systems is fed tylosin, rather than just those that get sick. Curiously, and also quite nicely for the farmers, tylosin also acts as a growth promotant, increasing feed conversion and growth rates during periods of use.

The problem with unnecessary antibiotic use, as we've been finding with over-prescription of antibiotics in humans, is the antibiotic resistance I mentioned earlier. The more frequently an antibiotic is used, the more quickly bacterial resistance can occur. And it's even worse for us if that antibiotic is one of those in a class used for humans.

Well, guess what? Tylosin – a growth promotant used not just in pigs, but in feedlot cattle and meat chickens – happens to be in a class of antibiotics, known as macrolides, that the World Health Organization has classified as 'critically important to humans'. In other words, dispensing it ad hoc to livestock on the basis the animals *might* otherwise get sick is a short-sighted, dangerous way to use one of the most valuable tools in the armoury of human health.

It's not just tylosin. You can add kitasamycin (another growth-promoting antibiotic) and oleandomycin to that list, both of which also belong to the macrolides class, and are used in animals. It's important to remember that once bacteria have developed resistance to one antibiotic, they are usually resistant to that whole class, or family, of antibiotics.

To counter antibiotic resistance, the considered response from the Australian Pesticides and Veterinary Medicines Authority (APVMA), the body charged with oversight of antibiotic use in livestock and pets, was simply to suggest a change in how antibiotics are labelled. 'Withdrawal of growth promotion claims from labels of macrolide antibiotics will form part of the solution to delay the emergence of antibiotic resistance,' the APVMA says helpfully, in their 2017 report. 'Combining this with the addition of a restraint statement on labels promoting prudent use of these products will assist in reducing resistance for these antibiotics.'

That's it. That's the veterinary medicine authority's entire response to the rise of antibiotic resistance. Just don't say an antibiotic is a growth promotant, and they'll have done their bit. They probably spent weeks refining those sentences under the watchful eyes of industry lobby groups. Their response seems like abrogating responsibility, if you ask me, because every pig farm *already knows* tylosin is a growth promotant. I would've thought banning prophylactic use altogether might be a more immediate and substantial change, and one that would deny tylosin's use as a de facto growth promoter. Using an antibiotic that promotes growth, on all the pigs in a farm, in the absence of disease, is an abominable use of technology that could potentially save human lives.

Not all antibiotics are critical for human health. Bacitracin, the most commonly sold antibiotic for animal use in the years 2005–2010, according to the APVMA (at around 30 per cent of all antibiotics sold), is also a decent growth promoter. Bacitracin is not as important in human health as macrolides such as tylosin and kitasamycin, though we may need to call on it again for topical (external) use if other antibiotics start to fail. In the same period, about 30 tonnes of antibiotics were sold every year in Australia as growth promotants (rather than for prophylaxis or therapeutics) – which is about 4–7 per cent of the total. We've been selling out a safer future, for the immediacy of cheap meat. I can't say this enough; it's just so Australians can have meat for dinner.

In *Big Farms Make Big Flu*, Wallace documents the many reasons for new and virulent disease outbreaks. One is antibiotic overuse. Another is intensive farming systems in which lots of animals are crammed together in close proximity, which also increases the need for antibiotics. Another factor is the speed at which an animal is grown: a faster-growing bird, for example, encourages a more virulent pathogen, because for a bug to be successful in its host's lifetime, it has to get a chance to multiply and spread to the next bird before its host dies. Also critical in the cycle are humidity (because bugs breed faster in warm, humid environments), and farm management practices, such as hygiene (things like faecal and litter management, waste water, and the cleaning schedules of the sheds). Wallace seems to be describing every modern chicken and pig farm.

———·———

Antibiotic resistance is a global issue. Looking overseas, one of the most high-profile cases of the widespread existence of livestock-induced bacterial resistance comes from pig farms in the United States. The bug in question is known colloquially as staph – a common bacteria that lives on our skin, but which has become resistant to the antibiotic methicillin (a relative of penicillin) that was rife in pig farming, giving rise to a strain known as methicillin-resistant *Staphylococcus aureus* (MRSA), usually called golden staph.

Staphylococcus, even some golden staph, is fine on our bodies, although it can cause boils and pimples and the like – but when it gets inside our bodies, through an open wound for example, or when we undergo surgery, it can become a huge problem, causing serious infections such as pneumonia and septicaemia. Golden staph, because of its resistance, has been a burden on hospitals for years. The longer you stay in hospital, the more likely you are to pick up golden staph. All antibiotics meet resistance, often within just a few years of being developed, but the more you use an antibiotic, the quicker the resistance spreads. The origins of MRSA are unclear, and undoubtedly overuse of antibiotics in humans is a major cause, but animals have had, and are still, having a role in its prevalence and increase.

So getting back to those pigs. Research from the United States pins an increased widespread emergence of the methicillin-resistant version of staph – MRSA – to pig farms, to the workers on the farms, and those who use the manure from pig farms to fertilise their paddocks. In 2009 a study showed 45 per cent of workers on American pig farms carried MRSA (are 'colonised' by it, is the term – as opposed to 'infected', which implies they are ill from it), and MRSA has also been noted in increased numbers in workers in pig farms in Europe. In Switzerland, pigs at slaughter that were contaminated with MRSA rose from about 2 per cent in 2008, to 20 per cent in 2012, a massive tenfold increase in just four years.

Now, these bugs, in this setting, aren't necessarily dangerous – but as with any resistant bug, when it gets in the wrong place, it can and has directly led to problems for those who work in those farms. While most of these so far have only been resistant infections in wounds, and pneumonia-like conditions in otherwise immune-compromised people, the fact that so many workers, and now people living in communities close to pig farms, have their nasal passages colonised by MRSA means a human health impact of some size is only a matter of time. In some reports, over 10 per cent of pork meat in Europe is also showing the presence of MRSA, where two decades ago it was virtually non-existent.

How does this happen? It turns out a derivative of penicillin – a vital antibiotic for human health – was being used in the pig feed on those farms. This alone is cause for concern because it leads to more widespread resistance (including to that whole class of antibiotics, the beta-lactams), but also, because the antibiotic isn't properly absorbed by the animal, 75 per cent of the antibiotics fed to them ends up in their manure. This is a problem for anybody in contact with the manure, including those using it as fertiliser.

Intensive farms are the perfect breeding ground for bugs, harmful ones included. And while bacteria are everywhere – in fact they account for up to 2–3 kilograms of our body weight, and we need them in order to survive – the ones we don't really want to be everywhere are everywhere at intensive farms, because manure is everywhere. Poo is in and on everything at intensive farms. That's the reason feedlots are rank, and why my clothes stank after visiting an intensive pig farm. And all that poo contains a lot of bugs that we probably don't want to get in the wrong place. If they're antibiotic-resistant bugs, that's even worse.

The pharmaceutical companies and some farmers may maintain that MRSA might have originated in hospitals, and that heaps of people who have contracted it have never been to an intensive farm. But the point here is that having more superbugs – as these antibiotic-resistant bacteria are called when they become resistant to more than one antibiotic – in the population can't be a good thing.

We know the United States has MRSA in pig farms. So does Europe. But the big unknown is China, where data are sparse, intensive farming has ramped up exponentially in recent years, and antibiotic use is rampant. There were about 700 million pigs in China in 2018, according to Reuters, so the potential for trouble is great, if not extraordinary.

The thing is, bacteria and viruses are always evolving, their super-fast breeding time meaning they can adapt incredibly quickly to any agents we use to treat them. The discovery of every antibiotic has been followed, often in remarkably short time, by resistant bacteria. It's nature's way.

It is how fast these resistant bacteria can spread that determines the size of the problem in the medium to long term. And antibiotic overuse, both in humans and in animal husbandry, is making that happen at ridiculous speed.

———·———

There's no hard and fast evidence of human disease getting far worse from the use of antibiotics on farms, but often that's because the research hasn't been done, or the pathway from animal feed to human disease has been hard to map.

But when US researchers looked at MRSA in pig farm workers, they found 64 per cent of them carried the resistant bug in their noses. They also found MRSA floating in the air 250 metres outside a pig farm, when you'd expect it to be bound to something more solid. According to a 2016 report in *Scientific American*:

> *researchers at the Johns Hopkins Bloomberg School of Public Health drove cars, windows down, behind trucks that were transporting chickens … they found antibiotic-resistant enterococci – a group of bacteria that causes 20,000 infections in the U.S. every year – in the air inside the cars, as well as resting on the top of soda cans in the cars' cup holder.*

(The study authors drove *behind* the trucks because, as independent researchers, they weren't allowed inside the actual sheds.)

The *Scientific American* article references a study by Stuart Levy, from way back in 1975. The research was funded by the Animal Health Institute and published in a 1976 edition of *The New England Journal of Medicine*. In it, Levy and his team fed low doses of tetracycline (a broad-spectrum antibiotic) to 150 previously antibiotic-free chickens, and had 150 control chickens as well, to see what happened. Within a week, almost all the *E. coli* (short for *Escherichia coli*) bacteria in the antibiotic-fed chickens' intestines were tetracycline-resistant. Three months later, the chicken's internal bacteria were also resistant to four other types of antibiotics. A month after that, the other 150 control chickens showed signs of resistant bacteria, and Levy and his team also proved that the resistant bacteria had begun to spread to the workers. It was the first time proof was found that low-dose, sub-therapeutic amounts of antibiotics, used as a growth promoter, could cause resistance.

Through years of other research on resistance, Levy became a powerful advocate for ridding animal feed of antibiotics. But it seems he was pretty much alone. In a profile article in *The Scientist* in 2015, Levy is quoted as saying: 'Our study from 1976 was the only prospective U.S. study on this, and industry didn't want more studies.' What's more, in 2015 Levy still reckoned nobody had tried to replicate his study to that day.

That's a 40-year lag. Four decades ago Levy had already worked out that bacteria might be able to pass on resistant genes to other completely novel bacteria, and that the resistant bacteria were quickly spread from animals to humans and back to other animals. You'd think someone would want to get better answers on what this meant for the future, but science goes a bit quiet in the intervening years – particularly up until the late 1990s, when antibiotic-resistant bugs in the general human population started to become a reality.

The reports I've read don't say, unequivocally, that antibiotic use on Australian farms has caused human disease, or increased antibiotic resistance in the general population.

But stay with me, as this next bit can get a little complicated. Trust me, I've spent way too many hours submerged in scientific speak about antibiotic resistance, and it's worth the trouble.

Here's what the Australian Pesticides and Veterinary Medicines Authority (APVMA) has to say in its latest review of tylosin and related antibiotics. I include the quote here because it means a lot to those in the trade, but skip past it if you want the translation.

> *Co-location of tylosin resistance determinants also allows co-selection of resistance to unrelated antimicrobial classes such as tetracyclines, -lactams, aminoglycosides and sulphonamides. Tylosin resistance determinants (and any co-located genes) are carried on*

a variety of mobile genetic elements and horizontal gene transfer allows wide dissemination of these genes to a range of related and unrelated bacterial genera. Importantly resistant zoonotic organisms carried by animals treated with tylosin at therapeutic or sub-therapeutic levels transfer to humans via the food chain (enteric organisms such as Campylobacter spp or enterococci) or direct contact with pigs, carcasses or pork (eg Streptococcus suis) and resistance determinants carried by these resistant strains can spread to other human pathogens. Apart from the direct effect of selecting for and maintaining resistant bacteria, the role of sub-therapeutic use of antimicrobials in increasing mutation rates and horizontal transfer of resistance genes must also be considered.

I'll try to decode that for those of us not in the industry. If you use tylosin and it creates a resistant bug, that bug may also be resistant to other, even more vital, classes of antibiotics. And the resistant gene, through a miracle of evolution that happens with all bacteria, can be handed over to completely other species of bacteria – including more dangerous bacteria. Then, the resistant pathogens (bugs) can be passed to us humans at the live stage of the pig and through the meat stage. The last bit says that using tylosin at all does this, but using it as a growth promotant does it more widely, and can have untold and unforeseen consequences.

And so, what was the response from the APVMA – the same mob that wrote this statement, and the same mob charged with regulating animal antibiotic use in Australia? Did they read their own conclusion and ban the prophylactic use of tylosin?

Nope. They said – as we saw earlier – let's change the labelling of tylosin so that it doesn't say 'growth promotant' anymore. No banning of its prophylactic use. Just change the label.

———·———

Levy's study was a warning, one the animal industry has chosen to ignore. Over the years they have taken advantage of different classes of antibiotics, ostensibly to reduce the risk of disease – but really with the aim of increasing weight gain, while also, maybe, reducing disease in the process.

The intensive animal industry is reliant on antibiotics because it is trying to defy nature. Farms where poo is so omnipresent that they 'smell like money', where the bacteria that line the gut of the animals get onto everything, and everyone, and *in* everyone, and where animals are packed into unnaturally high densities – these are farms that are breeding grounds for disease. That the drugs that might keep their animals alive are also helping get animals bigger faster has been a massive boon for the industry.

The APVMA believes that restraining antibiotic use will only delay the inevitable, and here I do agree with them. Antibiotic overuse in humans has played a possibly bigger part so far on human health. But with 70 per cent of the world's antibiotics now going to livestock, we are speeding towards ever more dangerous antibiotic resistance, and in the process squandering some of the most effective tools at our disposal against diseases that were once easily treatable.

The fact that we humans are putting ourselves in harm's way, intentionally, by the prophylactic use of antibiotics – antibiotics that are known for their ability to promote growth, hence their widespread use – is a blight on our nature. And I don't think future generations are going to thank us for it.

CHAPTER 14

Banned in 160 Nations – But Not Here

For a decade now I've had pigs. At first I just fattened two at the small farm I once owned. Now, the bigger farm we live and work on is named after the animals we first fell for. Fat Pig Farm is actually a mixed farm, where pigs only take up about 2 per cent of the land at any one time, but they make up an important part of what we are about.

Our farm doesn't resemble your average pig farm. That's a good thing, I think, after what I have learned on my travels. Remember the slaughter of mother pigs on the day they are due to give birth, to produce 'minimal-disease pigs'? But Australian pig farmers also have other practices they'd rather weren't too well known – secrets they generally don't share with their customers. I'll get to that in a minute, but first I want to describe our set-up, to be upfront with my biases and to help put other farms, and farming methods, in context. It might get a bit technical, but I think it's important, so stay with me.

After all, this is food we're talking about, something you're going to eat, and should, by definition, trust.

When you measure pig farms, you measure them in terms of the number of sows, the mother pigs. Australia's biggest piggery, Rivalea, has about 45,000 sows. Fat Pig Farm has Jackie, Audrey and Evita. So we're not really a very big or very usual pig farm. The pigs we rear – nicknamed by some in the industry as 'Commercial Suicide' – are a rare breed, the Wessex Saddleback. They are slower growing and have smaller litters, averaging about eight live piglets compared to the intensive industry's 12 or more. They are black with a white shoulder band – meaning they have a five o'clock shadow on the skin, which is considered unsightly. They also put on a decent layer of fat (hence our farm's name). All of which means they aren't welcome in the pork trade. Commercial suicide, they reckon, to rear them.

Our sows, all three of them, are given long grass, a shelter, and privacy to give birth in, away from the boar and other sows. They always build a large nest, akin to a haystack, in the 24 hours prior to giving birth (known as farrowing, in the trade). This nest they mostly choose to build within their shelter, though occasionally not. The piglets are free to roam for their entire lives, mostly on pasture, albeit inside certain fence lines, up until the time they go to the abattoir. We don't give them growth promotants, antibiotics (unless they fall ill), or other additives. We don't dock their tails (which is done in big crowded sheds to avoid tail biting), we don't cut their teeth (done to avoid fighting), and we don't ring their noses (done to avoid digging). We pretty much let nature take its course, from the birth to the weaning and beyond, allowing the pig to do what it feels it ought, only becoming involved in the process when necessary.

Pigs, for me as a city-raised boy, were my awakening about animal welfare. As intelligent as a kelpie or a young human, they're capable of higher thought. Sociable animals, they crave company and form close relationships, particularly with their brothers and sisters. A pig wants to spend about 65 per cent of its waking day digging in the earth, rooting around, exploring. The rest of the time is playfulness, or arguments, or rest. It doesn't matter if you feed a pig as much as it can eat, it will still spend hours each day digging around in the dirt. It's what a pig wants to do.

Mating happens naturally thanks to Barry, the hulking, well-tusked boar, the sows cycling every three weeks, starting about 4–5 days after being weaned.

Pigs, thanks to their want and need to dig the earth, along with the high winter rainfall we get on our farm, can damage the soil. They can dig a quarter acre over in a couple of weeks when the ground is soft. For that reason we spend a lot of time moving pigs to new grass. We reckon it takes about ten times as much labour to rear pigs our way, and about twice as much feed compared to shedded pigs. We are, undoubtedly, at the most inefficient end of the spectrum.

———

I describe all this because it gives context to how pig farming can be done – but only in a small way, and only to produce a very tiny amount of the pork Australians want to eat. Free-range pork constitutes a fraction of the market, at most about 5 per cent. Pasture-raised pork (where pigs are put on fresh grass, not just stuck in a pig sty) is an even tinier fraction of the total. When you consider that over 70 per cent of the pork we eat in Australia comes from overseas, where conditions are possibly worse than here, what pastured pork producers do rear is a rare and insignificant portion of the whole.

The vast majority of pigs in Australia come from a system in which they don't get to express their instincts. Most sows are trapped inside a shed, some on slatted floors, some of them on concrete, and about 20–30 per cent of them are still in sow stalls as I write this. A sow stall is the bath tub–sized cage that doesn't allow the pig to do anything except stand in pretty much one spot, and lie down, with no ability to turn around or walk during their whole 115-day pregnancy. A voluntary industry phase-out brought in a few years ago has (ahem) stalled – although thanks to consumer concerns (and the buying power of Coles, one of Australia's two big supermarkets), at least the majority of sows are now able to move around a little bit in group housing, rather than stalls.

Sow stalls weren't invented to be cruel to pigs. I don't believe so, anyway. They were invented because when pregnant pigs come in contact with each other, they fight. Like *really* fight. It's terrifying to see sows weighing 150 kilograms or more charging, biting, attacking each other. There's blood, gouging, torn ears and tails. It really is a battle. To fix the problem, farmers began locking the pigs in cages so they couldn't fight. To the farmer it was about *improving* animal welfare. Sow stalls helped to avoid fear, pain and injury.

Farrowing crates, a similarly restrictive device used when the sow gives birth and until weaning, were also invented to fix a problem. Nobody thought, 'Oh goody, we can lock pigs up in cages while they give birth and suckle their young, just for kicks!' No, these cages were invented because if 150 kilos of mother pig suddenly lies down, and a dozen or more barely-out-of-the-womb piglets don't move out of the way fast enough, they can get crushed.

But both the farrowing crate and the sow stall are the result of reductive thinking. Baby pigs can still get crushed in sheds, albeit at a lower rate, despite the use of farrowing crates. And baby pigs get crushed in the wild, too, where feral sows give birth to much smaller litters. The issue comes with what we perceive to be priorities, and the way we solve problems when not taking the animal's emotional and intellectual wellbeing into account. For both of these problems – sows fighting and piglets crushed – we've come up with solutions that suit us, but that don't suit the pig in all except one measurable way: preventing fighting or crushing, depending on the cage. It's not about what the *pig* wants, but what the farmer wants.

You can, if you desire, breed for less crushing by pigs. Mothering instinct can actually be bred for, if that's where we put our efforts. Sadly we don't. We breed for colour, leanness, big litters, fast growth, and the ability to be confined. Fighting sows can be managed, as most of Australia's intensive pig farms are now doing, simply by giving them more space, or letting them see each other through a fence for a while before moving them in together. On our farm, our sows can fight a bit when first put back together, despite having lived together for years on and off. But after a brief period of asserting dominance and establishing a pecking order, usually about a day, they settle into their roles amicably. For us, giving two sows about a quarter acre to run around in allows them to express their natural behaviours, not only in terms of nesting and foraging, but also allowing less dominant pigs space to simply run away.

Piglets dying is not something I expected when I first got into farming. It's a damn hard thing to accept. But it's important to remember that however sad, heartbreakingly sad, as these crushings are, in nature pigs have lots of piglets over their lifetimes because most of the little ones don't make it. In order to replace themselves, sows have to give birth to numerous young.

Farmers put conditions in place where domesticated animals have a far better chance of rearing live young to breeding age. We're really good at it. Nature would take out most of the young well before a farmer sends their fattened pigs to the cutting shop. But

I'm not sure the farrowing crate meets the needs of the mothers who are trapped in them.

———·———

The modern industrial pig is very different from the old breeds, such as Gloucestershire Old Spots, Oxford Sandy and Black, Wessex Saddleback, Large Black, and Tamworth to name a few. Big companies own the genetics to the very long, white pigs used in the industry – a variation on the Danish Landrace. They lease boars and sows from known genetic stock to smaller farmers who are licensed to breed from them, quite often the Landrace cross: a pig bred to have chunky thighs (for more ham), lean (because we all got warned off animal fat, and onto margarine, in the 1970s), fast growing (for efficiency), with large litters (ditto), and able to cope with the mind-numbing confinement of the sow stall and farrowing crate.

Modern industrial pigs are always white, because you don't skin a pig like you do cattle and sheep. They're scalded after being killed, and before being gutted, by being dunked in a hot water bath and scraped until the hair comes off. The roots of black or ginger hairs on older, coloured breeds can leave a five o'clock shadow in the rind – an anathema to the retailer, as consumers simply don't like the look of it. (For the same reason, every commercial meat chicken and turkey is white feathered.) The truth is, when you eat pork that has the rind on, or has the rind cooked with it, you generally eat the stubble of hair – it's just that on a white pig you can't see it.

Each litter of piglets born to the modern sow is about three times bigger in number than her wild forebears. She has to have 14 teats, to suckle so many young. Her body is remarkably large and elongated, as are her children, because everybody loves bacon, and that comes from the middle of a pig. A modern pig, kept relatively inactive indoors compared to my farm's hard-digging, hard-running pigs, can reach the same weight in close to half the time, while putting on remarkably little fat. They are, in so many respects, a far more efficient pig when you look at lean weight gain alone.

Now, there's lots of reasons we have the 'Commercial Suicide' breed on our farm. An older breed, and one of the first brought into Australia, the Wessex Saddleback is a hardy pig that doesn't cope well with confinement, but does cope well with our climate. After all, it's originally from the United Kingdom, so Tasmania is like the tropics to them – drier and warmer than the old country. Our pigs have an average of eight live young, well below the industry average of 12–13 for Landrace pigs. The industry best practice is for each sow to have 2.4 litters a year, whereas we only have about two at most. Partly that's because we wean much later, at about 70 days rather than the industry standard of about 25 days.

So we have these pigs because they suit our farming system (pastured, free-ranging), and because they can handle our conditions, but mostly because there's only one reason the pigs are here, really, and that's because we are going to eat them. We have these pigs because they taste nice.

Industrial pigs have lots of advantages, as we have seen. They get bigger quicker, are obviously way more efficient all the way through their life cycle, and they're leaner. But the leaner thing, and the efficiency thing – well, these characteristics are also both products of the pharmaceutical industry. Modern pork is more the result of chemical inputs than it is of nature.

If you wean pigs around the time the industry does, some from around two weeks old, others from about three weeks, piglets die. Often it's from gut inflammations or similar, possibly because even at the industry standard age of 25 days old, that's still quite a young age to wean a mammal. To prevent this, the entire litter are fed antibiotics, as a prophylactic (a disease preventative). On one farm I visited, the piglets were fed antibiotics from five days prior to weaning (which occurs around the two-week mark), and for five or so weeks after weaning. While this does increase growth rates, as we saw in the last chapter, this practice isn't a selling point in terms of public health.

To finish the pigs, they're often fed one or another of two other chemicals, which also act as growth promotants. These chemicals also reduce the fat that we prize so much on our little farm. Paylean – the brand name for the feed additive ractopamine hydrochloride – is banned in about 160 countries because of the adverse reactions on the living pig. But it's not banned here in Australia, and is routinely used on finishing pigs. It makes the animals put on lean muscle mass, and helps strip them of fat. It helps defy the natural body structure of the pig.

Australian pigs are also regularly fed Reporcin, a brand of porcine somatotropin (pST), or pig growth hormone. According to a submission by Australian Pork Limited to the federal government:

> It is administered to finishing pigs when their natural
> levels of pST are in decline to ensure that they continue
> to deposit muscle tissue and reduce the amount of fat in
> the carcass as required by consumers.

Note the words, 'as required by consumers'. So it's *your* fault they add hormones to pig food.

———·———

By this stage in my research I was wondering if pigs are just a chemical cocktail dreamt up in a lab. Then I learned about Improvac, the chemical castrator, which according to the maker, Pfizer, can lead to about 4 kilograms extra lean muscle mass at slaughter, along with an overall larger pig for the same feed in the same time. (Improvac's original aim, to reduce the hormonal 'boar taint' that can occur in male pigs after reaching puberty, is a more noble goal. But hey, Pfizer wants to sell the product, so why not tout its growth-promoting qualities, too?)

Now, all of this chemical input, I'm sure it's fine. I'm sure the Australian Pesticides and Veterinary Medicines Authority, who we met earlier in the previous chapter on antibiotics, are all over it. Some of this chemical input reduces the need for feed and so, it could be argued, reduces the ecological footprint of raising pigs. The problem, from my point of view – someone who wants pork to taste good – is that in order to make the pork lean, they actually make it tough and dry.

For the 20,000 years or so that we've been domesticating pigs, we've included flavour as one of the major characteristics we care about. We've bred for various features in domesticated animals. For size, for bigger litters, for domesticity so they don't eat us. For health. For flavour, because that's quite important when you think about the end use of the product: food. For about the last 80 years, though, flavour – along with mothering instinct – hasn't really been on the agenda. So these fast-growing, relatively immobile pigs, who put on lean muscle mass thanks to some work by research chemists, are a bit dry and tough.

That's why the industry has come up with a solution: moisture-enhanced or moisture-infused pork, where it's pumped full of salty water to make up for the lack of flavour, and the lack of fat – with the beneficial side effect of making the meat heavier, so instead of just selling you meat, they're also selling you water.

Don't get me wrong. While we do a bunch of stuff on Australian pig farms that I don't like, maybe those things would meet community expectations – if the community knew about them. But thanks to a less than open and accountable industry, most in the community aren't aware of everything that is being done in their name.

Andrew Spencer, Australian Pork Limited's CEO who is due to step down as this book goes to press, stated in the *Australian Pork Newspaper* in March 2016 – about the time I was trying to get access to our most intensive pig farms – that they 'are not afraid of others seeing what we do'.

Well, I'm afraid Andrew, you were. Your farms still are. Your farmers are. Your vets are, and the mostly large companies that dominate the industry in Australia are. The pig industry *wants* us to believe that they do things well, that they do things sustainably, that they do things that the public would not only countenance, but also condone.

And yet, at the same time, they don't *want* the public to see how things are actually done. They just want you to eat dinner and stay none the wiser.

I know I'm possibly sounding like a luddite, thinking the chemicals that we use to help modern pigs grow are unnecessary and possibly detrimental to the eating quality of the pork these pigs produce. And of course my views are tainted by my predilection for more natural behaviours, more natural growth. But I also believe that meat eaters should be given the information on how their meat is reared, and what tricks are used to make it grow differently or faster.

To help unearth just what they're fed, I've been digging around, and since you're still with me, it's also worth looking at what most industrial pigs get to eat.

Following are the ingredients in a pig feed, called Grolean, available in the United States – the same brand name promoted by

Rivalea, Australia's largest piggery, on their website (though because Rivalea have chosen not to engage with me, I can't be sure that it's the same formulation used here). It's the feed for grower pigs.

> *Grain Products, Processed Grain Byproducts, Plant Protein Products, Calcium Salts of Long Chain Fatty Acids, Monocalcium Phosphate, Dicalcium Phosphate, Salt, Calcium Carbonate, Vitamin A Supplement, Vitamin D3 Supplement, Vitamin E Supplement, Yucca Schidigera Extract, Mineral Oil, Niacin Supplement, Vitamin B-12 Supplement, Calcium Pantothenate, Riboflavin Supplement, Menadione Sodium Bisulfite Complex (source of vitamin K activity), L-Lysine Hydrochloride, Sodium Selenite, Ferrous Sulfate, Zinc Sulfate, Copper Sulfate, Manganese Sulfate, Ethylenediamine Dihydriodide, DL-Methionine, L-Threonine, Dried Aspergillus Niger Fermentation Product.*

Grolean also contains tylosin, the antibiotic we met earlier, as the 'active drug ingredient' at 20 grams/ton, and they make quite clear that the feed can't be used as a growth promotant – even though its formulation remains unchanged after new US and Australian regulations forbade that term on the label in 2017. (It's all about perceptions.)

In other words, a vet only has to say you can use the feed – just not on the basis of growth promotion. You can still use it if the vet lets you. I don't know if that's the same formulation as used in Australia – but remember, even if isn't, over 70 per cent of the pork we eat is imported, much of it from the United States, where this feed is used.

—·—

Now, big piggeries, and even some small boutique free-range pig farms, will tell you that they've hired nutritionists to formulate their feed – hence a list like the above. I've been questioned on why we don't have a nutritionist work with us.

Because, well, forgive me for a little cynicism, but aren't they the people who persuaded us all to give up butter in favour of margarine in the 1970s? Aren't they the people who, despite untold amounts of research and huge financial rewards, can't yet make a formula that is a complete and satisfactory replacement for breast milk? Aren't they also the kind of people who suggest a balanced diet of leafy greens, vegetables, grains and possibly small amounts of meat for us, yet come up with this litany of strange additives for an animal that has successfully hung around farmhouses for the last 10,000 years, eating leftovers while also breeding and being fattened for the pot? Do we really need a nutritionist to tell us what a pig probably should be eating, if their gut is very similar to ours, and they've thrived in the wild since forever? And if nutritionists tell us pigs should be eating 'Menadione Sodium Bisulfite Complex' and 'Dried Aspergillus Niger Fermentation Product', do we really think they've got the pig's best interests at heart? Or is it the feed company's?

All of this, just to create your next pork chop?

Pig farming, as we do it in Australia, is flawed. But overseas is possibly worse. As we've seen, most of the boneless pork that is consumed in Australia – 70 per cent of the total pork in the nation – is imported. Much of that comes from systems that are worse than ours. Increasingly it's coming from China, where half the world's pigs are kept. In 2013, 16,000 dead pigs were found in the Huangpu River, which flows through Shanghai. They were sick pigs, presumably dumped by pig farms upriver to simply rid themselves of a problem. In the United States, at least until recently, growing pigs could be fed not just tylosin, but carbadox, a carcinogen that induced birth defects in animals, including in humans. (The United States is in the process of withdrawing carbadox from swine feed as I write.)

———·———

Despite Australian Pork Limited's protestations to the contrary, what pig farmers do around Australia – and in China and the United States – isn't done in the public eye. I wonder if our own industry might not meet consumer expectations around animal husbandry and chemical inputs? And if they do, then surely it'd be a marketing bonus to promote what happens in our own backyard?

I know I'm biased about pigs, because I really, really do like them. I really, really like the way they can recycle waste and turn what we can't or won't eat into high-quality protein. I really like the fat we get from them that we can flavour our vegetables or pulses or flaky pastry with. And I really do question whether the thinking eater in Australia can eat pork with a clear conscience – even Australian pork, which is probably at least one step above the imported stuff – unless the industry really cleans up its act.

This is an industry that has thrived on secrecy. An industry that has ignored the need for social licence by remaining aloof to community standards and accountability. It's also an industry on the verge of collapse thanks to cheap imports, and the public's lack of confidence that the local farms are doing anything better or more ethical than pig producers overseas.

I have strong opinions on pigs that I know won't necessarily translate to general community standards. But I'm proud to have standards. I don't like the unquestioning use of antibiotics around weaning age. I don't like the 28-day incarceration of pigs in farrowing pens when there are alternatives, even for intensive farms. I certainly don't like 'snatch farrowing' and 'minimal-disease pigs', where the babies are either snatched from the mother at birth, or the mother slaughtered at birth.

I'd also like to think pigs aren't forced to grow quicker and leaner with chemicals such as ractopamine, porcine somatotropin and Improvac, and can be fed something close to a natural diet, and still thrive and possibly taste better.

But it doesn't matter what I think, because I don't buy pork from that system to feed my family. What matters is what you, the general public, think – and how your shopping will change because of it.

CHAPTER 15

What's Behind the Locked Gates?

One of my main objectives when researching how we produce meat, both for my work in print and on television, has been to get behind the locked gates of Australia's most intensive farms.

Despite my own biases and beliefs in our own farming system at Fat Pig Farm, with its extensive, rather intensive animal husbandry, I'm not sure that what we and others like us do will be enough to feed the world. More importantly, I'm interested in how meat is raised for the general population – not just the few; the wealthy few. It would seem odd to me that what is done for Australians, on Australian soil, for Australian dinner tables, could be such a big secret that we, the public, can't be entrusted with knowledge of it. After all, if you're really a proud chicken farmer, wouldn't you want the public to know you have not only the birds', but the customers', best interests at heart?

Maybe I'm naïve. Or at least I once was.

Between myself and people working on my behalf, we spent more than a year trying to get me onto Australia's most intensive farms. My particular focus was on the two industries that thrive on rural imagery, and promote themselves as more sustainable, with lower greenhouse gas emissions: the pork and poultry farms. Their peak bodies are Australian Pork Limited, and the Australian Chicken Meat Federation. And both in my view are as slippery as a well-oiled piglet.

After several weeks of back and forth, Australian Pork Limited did line up an appointment to chat to their CEO at the time, Andrew Spencer. He gave me interesting statistics, particularly on imported meat – which makes up about 70 per cent of the boneless pork meat in Australia, most of which is used for smallgoods (including ham and bacon), and which comes from systems we have no control over. However, he wouldn't let us film the meeting, which means we couldn't use any of the material in our series, *For the Love of Meat*. He told me that virtually all the pork meat produced in Australia comes from about five major companies, of which two dominate the market. But he wasn't really helpful in terms of access.

Less helpful still was the Australian Chicken Meat Federation, who years ago promised a farm visit and a chat with their CEO. A chat that never eventuated.

I pursued them more recently, over a period of months, reminding them they had offered us a visit to one of their intensive farms. I was thinking, after they'd suggested I visit a farm a few years ago, that television was an excellent way to 'dispel some myths', as they put it, around chicken farming. Letting people see what you do is a really good way to gain the trust of the public. And after months of correspondence, and numerous calls, emails and even texts, and following repeated attempts to get onto a chicken farm, we were fobbed off with talk about biosecurity, future dates (possibly after you've finished filming, sadly ...), and the problem with cameras, bacteria etc. The new CEO, Dr Vivien Kite, never did return a call or email, choosing instead to go through a media firm, and to date never fulfilling the previous CEO's promise of a one-on-one chat. On or off camera.

The Australian Chicken Meat Federation, to be fair, isn't there to show us what you and I want to know. They represent the chicken meat companies, only seven of which produce virtually all the chicken meat for sale in Australia. The two largest companies – Baiada (under the brands Lilydale and Steggles) and Ingham's (which also uses the brand Marion Bay Free Range) – sell about 70 per cent of the nation's total chicken production. About 430 million birds a year.

From my perspective, I thought it'd be good for the chicken industry to tell their own story, rather than let animal rights activists tell it for them, or their publicity firms. After all, 650 million chickens don't die so chicken farmers can eat them – they are reared in the name of their *customers*, the general population of Australia. While we pursued access (and while it was increasingly looking like we'd fail), we conducted an experiment on our farm to demonstrate the stocking densities of the most intensive sheds.

We set up part of one of our farm buildings as an intensive chicken farm, complete with deep litter, using the guidelines for a non-mechanically ventilated shed for stocking densities. (It's a bit technical, but simply put you can have more birds in a better ventilated shed.) We only housed about two-thirds of the birds that would be allowed under these guidelines, raked the litter (a mix of sawdust and chaff) every day, and gave the hatchlings places to perch and close proximity to food and water. But even at stocking densities of far less than the industry's 28 kilograms of bird per square metre for an unventilated shed, the experiment was abandoned when I couldn't stand the animal welfare issues any longer. Well before that, we'd already destocked by moving 20 per cent of the birds to another farm.

It was a flawed experiment, but telling nonetheless. Even in a few short weeks, we ended up with birds that showed signs of hock burn and breast blisters, lesions that come from constant sitting on the floor, the build-up of ammonia from their excrement on the floor, and not exercising enough. Most of the birds put on so much weight they outgrew their body's ability to move much. In the same shed we had two types of birds: the usual white commercial genetic stock, and a variety bred to free range, Sommerlad chickens. The Sommerlads, interestingly, showed no such lesions, were more alert, more robust. But still, it wasn't a real shed, just our interpretation of one, so I was pretty keen to actually get into the most intensive sheds in the country to see what their birds looked like.

Now, we bent over backwards to get into chicken sheds, to get the Australian Chicken Meat Federation on board to allow us into farms. They could use their own camera operator. We could use a GoPro, a waterproof camera that could be doused in toxic amounts of bleach prior to entry. Zero biosecurity risk. But still, no filming.

From my experience, the chicken federation's main aim seems to be to disguise how chicken is produced, and to fob off journalists interested in showing the public a modern chicken shed, while getting you to believe in their fairytale images and eat more chook. Sadly, this strategy works. They did everything they could to get us to just fade away. *Nothing to see here*, is the attitude. A strategy that's worked for years and years.

To circumvent the Australian Chicken Meat Federation and their stalling techniques, we started to approach individual chicken farms to see if they were proud enough of what they did to show the public. The first to say yes, as you can imagine, wasn't the most intensive one. It was an organic producer in Queensland, Inglewood Farms, producing Inglewood Organic poultry. They allow their chooks to range outside, they grow their own organic feed for them, and slaughter them on site. We were welcome, very welcome, to see inside the sheds, and it looked like a state-of-the-art facility, with the birds able to come and go as they pleased during daylight. It didn't look like a farmhouse from the 1800s – rather, a modern, intensive, but not confronting place that reared a lot of birds that were able to pursue their dust-bathing, grass-pecking desires. Stocking densities are lower for organically reared chickens. Along with the extra ability to range on grass outside, the space inside the shed is about 30 per cent greater per bird. The animals themselves looked like a bunch of standard meat-breed white birds (of the same style we had reared at our farm), which could, in many ways, express their instincts by being able to perch and forage and range.

It showed that chicken farmers can be proud of what they do, and happy to show consumers what they do, and I think with good reason. Despite the heat (we went to Queensland in summer), Inglewood Farms didn't reek of chicken poo, and everything about the place seemed to consider the environment, the birds, and the workers.

The next farm to allow us access was an RSPCA-approved shed. The standards set by the RSPCA (Royal Society for the Prevention of Cruelty to Animals) cop a lot of criticism from those interested in really high-welfare chickens. In the basic standard, RSPCA-approved birds are shed-bound their whole lives, so they can't necessarily free range – but they do have a lower stocking density than in standard sheds. The rules are stricter on the amount and cleanliness of their litter (the sawdust etc. under the birds' feet); on having to provide perches to increase leg strength; and being required to provide a few items (such as hanging CDs) for mental and physical stimulation. Importantly, RSPCA-approved sheds must also have periods of full dark and full light so the birds can go to sleep, and also fully wake up.

While the RSPCA standard isn't exactly a gold standard, the view is that by decreasing leg breakages, by having dry litter to reduce health problems such as hock burn and breast blisters, then the lives of millions of birds have been improved. And it was only through the actions of the RSPCA – not the Australian Chicken Meat Federation – that we were allowed in the shed at all. Also, the company did allow us to film; I talk about the visit in Chapter 4 (160,000 Deaths: A Medium-Sized Operation).

The last farm we managed to film at, however, goes to the heart of the problem with intensive animal production. A certified free-range farm, they'd originally said no to filming, which struck me as odd; usually the higher-welfare farms say yes (like Inglewood Farms), and free range obviously seems aimed at higher than average welfare standards. But it was only through Guzman y Gomez – a fast-food chain that has committed to using free-range birds under the Lilydale brand – that we managed to snaffle a visit to this one farm. It took months of organising, weeks of negotiating terms, and some fairly serious sucking up to the parent company, Baiada, before we were granted access. Every day leading up to the visit, it looked like it would fall through, with more and more conditions placed on us as time went on.

On the day of filming, with the head of Guzman y Gomez excitedly pointing to chickens running from the sheds onto the shrubbed, grassed area outside the pens just after dawn, I felt Baiada's secrecy had been misplaced. This was well-managed intensive free-range farming, albeit using the same relatively flavourless breed of bird that is ubiquitously used. The system had fixed sheds, meaning

– like the organic farm – that some areas were less grassed than others, but the birds had space to dust-bathe and seemed to revel in their capacity to roam.

It was only when we wanted to look *inside* the sheds – the space where the birds spend the night and much of their day, and where their food and water is kept – that things went south. There was no way Baiada wanted a film crew to see it.

Despite assuring them of no evil intent and suggesting they'd look better on telly if they actually let us see how things operated in the shed, the company's owners, publicity people and managers argued the case. I said I wanted to show Australian consumers what an intensive free-range farm looked like and all hell broke loose. 'We don't call it *intensive*,' they argued, and threatened to kick us off the property.

After months of trying to placate chicken farmers, of being ignored by the Australian Chicken Meat Federation, I was tired of tiptoeing around everybody else's sensibilities. How is raising tens of thousands of birds in a shed not intensive? On what planet do the chicken farm owners and workers live, where this isn't a suitable term? Was there something in the sheds they were embarrassed about? Did they not pick up all the dead birds after the overnight rest, something I know would've happened on our other farm visits? (The mortality rate for chooks can be some 4–12 per cent, so if you have 50,000 birds in a shed, averaged over 6–7 weeks, each morning there's potentially 40 or so birds that have carked it the previous day.)

Whatever the reason, they arced up and wouldn't let us stay until I apologised – and even then, they still only let me see inside the shed, and not take the camera crew.

And you know what I saw? Pretty much exactly what I'd seen on the RSPCA-approved farm, with slightly fewer birds because so many were outside. The same automated feeders. The same style of drippers for water. The same white birds, albeit it a little more alert than their shed-bound cousins.

(Just a side note: if you look up the Australian Chicken Meat Federation's code of practice about antibiotic stewardship, they refer to 'intensive chicken production' as a matter of course. The farmers know it's intensive, they just prefer to use the word 'farm' in their labelling and publicity, which invokes storybook imagery and doesn't really define the style very well at all.)

Perhaps I'd been naïve. When we visited this chicken shed, one that boasted it was free range, I had assumed the owners would be proud of what they did.

The problem for Baiada – and the reason they don't like the term 'intensive', even if they have tens of thousands of birds in a shed – is that 85–90 per cent of their chickens *aren't* free range. The same company that has free-range birds in this setting also has *way* more birds that will never be let out of their sheds, and whose only experience of daylight is probably as the sun rises at the slaughterhouse. I reckon they thought that by exposing what they do in the better free-range sheds, we'd be undercutting what makes their business tick: all those far more intensively stocked sheds with birds that can't range free. The vast *majority* of their sheds. The ones they DEFINITELY would never let us film. (The same could probably be said of Ingham's, who own the Marion Bay free-range brand, and whose reticence to film may also have to do with their decidedly un-free-range Ingham's label.)

Why was this the case? Well, pondering upon it later, I think it's all about what they – the owners and managers of these chicken farms – believe in. I don't think this free-range farm was there because of welfare principles, but because the owners know there is a market for free-range hens. Nine out of ten chickens they produce are in the more highly stocked, more intensively reared, non-free-range farms. They only had the higher-welfare birds because the market had demanded something better, and because their customers (in this case Guzman y Gomez) expect it. It's encouraging to see consumer demands making a difference, but disheartening to think it doesn't come from some moral backbone from within the industry itself – just raw capitalism at work.

It looks nice for the industry as a whole to show a bunch of trees and grass with free-range birds just being birds. But, show a shed with, say, 40 kilograms of birds per square metre – which is what all Baiada's other conventional sheds would probably boast – and, well, the public might realise how bad this all looks for their cheaper, lower-welfare birds.

It certainly looked bad when Lilydale's parent company, Baiada, got clocked for deceptive conduct in 2013 for promoting its shedded Steggles brand chooks as 'free to roam'. According to a report in *The Australian* newspaper, Federal Court Judge Richard Tracey, who during the trial toured several chicken sheds in which the birds were supposed to be 'free to roam', found the amount of space for each chicken and their consequent ability to 'roam' varied significantly at different stages of their life cycle.

'Chickens could not, in my judgment, be said to be free to move around the sheds at will and with a sufficient degree of unimpeded movement to justify the assertion that they were free to roam,' Justice Tracey concluded.

In other words, Baiada marketed the chickens using terms they knew customers would favour, when not bothering to actually back up these terms by action on the ground. Justice Tracey went further:

> *They could not move more than a metre or so*
> *(at most) without having their further movement*
> *obstructed by a barrier of clustered birds.*

Continued movement could only occur if a bird pushed into the group or encountered a small space between other birds.

Justice Richard Tracey found the Australian Chicken Meat Federation, the peak poultry industry body, which should clearly know better, had also engaged in misleading conduct by saying such intensively shedded birds were 'free to roam' and had breached consumer law. Perhaps that's why Baiada and the chicken federation were so wary of us seeing inside the sheds. Justice Tracey had kind of done it for us, and what he saw, he didn't think lived up to their hype. It's a shame that along with his legal access he didn't take a television camera. It would be insightful to show consumers the sheds he visited, and what the reality of an average 17–19 birds per square metre actually looks like. Nevertheless, the intensively reared hens that weren't free to roam did cost Baiada $400,000 in fines.

So, I can tell you about the intensive poultry industry. It's evasive, secretive, obstructive and unaccountable. The word untrustworthy also springs to mind. Dishonest? I think the deceptive conduct rulings say that better than I can. The losers, apart from customers, are obviously the chickens, whose welfare doesn't appear to be high on most company's lists.

Now, I did make it into the best intensive poultry sheds in the country: the organic, the free range, the RSPCA-approved. But all of these still fall a long way short of the ideal farm we have in our heads – the one marketing people like to use, of a few chickens on a lot of grass. The Charlotte's Web of farms.

For those, I had to head to pastured chicken farms, and even some where different genetics produce resilient, slower-growing birds with inherent flavour, such as the Sommerlad breed mentioned earlier. These farms, which I'll touch on later in Chapter 24, actually look like the chicken farm from storybooks, albeit with way more birds, clever mobile shelters, with acres of grass and low stocking densities. What's more, they have the kinds of birds that don't all look the same white colour, and do give you joy as a cook.

So the best chicken farms are reminiscent of a storybook. The worst chicken farms? Well, they remain closed to me, closed to you. But you can imagine what they must look like. And smell like.

———·———

The same is true for pork. While Australian Pork Limited made the right noises to my face, behind the scenes they were less than helpful. I never made it into the biggest farms, the most intensive sheds, some of whom told us Australian Pork Limited had warned them off.

We were rebuffed by farms that – in their own carefully curated media releases and advertising – make big claims about how their sows live outdoors, or aren't kept in confinement anymore.

At a meeting with Rivalea, the company that runs the largest pig farm in Australia with about 45,000 sows, we were surprised to learn that its Singaporean owners had no interest in what the consumers of the pork actually want. They don't really care if you expect to know how Rivalea's half a million or more pigs are reared each year, because, well, they just don't. And so we weren't allowed to see the farms.

And all of this we found out by going through the front door – or at least trying to. The whole idea was to demystify meat for the omnivore. If you want to convince yourself of your veganism, it's easy to find footage of cruelty on the web. But I wanted to see what proud farmers can show us about how meat is reared for the price we're willing to pay. I wanted to see the confluence of human ingenuity and animal welfare, the modern farm that we could all be proud of.

But that door was consistently closed, and often locked or even double bolted.

I worry about an industry that does things they don't think the public would like the look of. I doubt the motivations of those who are not proud enough of what they do to let the story be told – and let's face it, they do it in everyone's name. I don't trust the intensive animal industry one little bit from my experience of their double speak, obfuscation, secrecy and what little information is leaked or been dragged out through the courts.

So if you can't get in through the front gate, should you go in through the back? This is what animal rights groups do. They break into people's farms, in the dead of night, cameras at the ready. With their pre-arranged agenda (to stop all people eating meat, usually), they try to find a badly managed farm, focus on the sickest animals, the feather-pecked chicken lying in the corner, or the just-crushed piglets, and give all farmers a bad name.

I chose not to join animal groups on their night raids, as a matter of principle, despite the offer being made. I don't like the idea of people breaking into my house at night, and can imagine the fear and violation that farmers who have had raids must feel. (And also, the fear the animals must feel with strangers coming at them in the night. There's also the threat of disease, too.) And I really don't think it should be up to animal activists to be policing animal welfare, when most of them don't want us to be farming animals at all.

I get why the activists don't wait to be invited in. I really do. Trying to get in the front door plays into the intensive farm lobby's hands, because they just keep promising and not delivering; keep paying lip service but not coming up with the goods. And if you do get in, it's to the industry's best practice farm.

But condemnation of everything to do with meat, because you're first and foremost of the view that *nobody* should eat meat, is a flawed strategy. It has certainly helped some animal welfare cases in the short term, but in the longer term, it has pushed the industry into even more secrecy. It has made sure 'biosecurity' and 'commercial in confidence' are words used to muddy the waters, where 'We don't want you to see what we do' would be more accurate.

Some of us do believe that there are good farmers, and dodgy farmers. There are good ways to rear animals, and not so good ways – and arguably, terrible ways.

The more that the intensive animal industry hides things, the more we don't trust them as farmers. The more they say 'Don't worry' and 'Nothing to see here', the less we believe them – especially when activists do break in and record footage of animal cruelty, footage that leaves a stain on all our souls.

Do I feel sorry for the poor farmer, who may have their own health issues, or depression, or relationship breakdowns to deal with? Things that impact their ability to always watch over the animals for which they care? Yes and no. The simple fact is that an industry that doesn't trust its own consumers with knowledge, and that insults the intelligence and nature of meat-eating consumers by denying them access to information about how their next meal was raised, probably deserves the pillory it receives.

But I also know there are good farmers out there, who are proud of what they do, and how they do it. They're more than happy to show us how they farm. To produce meat, something has to die, regardless of the farming system it lived in. Often we're sheltered through choice, but also through the divide between producer and consumer.

Surely an informed public is better than one that blindly buys their food from an anonymous system? I don't think there's anything wrong with asking where your meat came from. And I don't think it's unreasonable to expect an answer when you ask.

The farmers I hang around with are not afraid of accountability. They are not afraid of the tough questions about taking an animal's life so that we can eat meat. The farmers I admire are open about what they do, and can justify it to the end user, and some of those have thousands of pigs run in a way I may not approve of myself. Those farmers still want social licence. They want their workers to hold their heads high and be proud of what they do, knowing that the Australian population, as a whole, agrees with what is done on behalf of the public. These farmers aren't embarrassed or ashamed of how they farm, and are willing to let the nation see what they do, even if it means having to explain how modern farming systems work.

Sadly, the intensive chicken meat and pork industries in Australia don't, in general, behave like that. To many farmers' discredit, I've found the attitude of profit over social licence pervasive, as well as counterproductive.

I didn't get into the worst farms, and I do hope they aren't as bad as my imagination conjures up. But only if journalists are given more access, if farmers are truly proud of how they raise animals for meat, will we really know for sure.

CHAPTER 16

How Do We Feed the World?
Animals May Well Be Part of the Plan

Globally, hunger is a problem. Too many people not eating enough, or not eating enough of the things they need to be as healthy, active and able as possible. And with the world's population recently hitting 7.6 billion – and with the talk of the carrying capacity of the earth to feed no more than about 10 billion sustainably, a number we'll hit in the next few decades – we need to talk about food security.

One of the realities of growing food is that life really is a cycle. In nature, plants and animals symbiotically co-exist, the animals eating the plants, the manure from animals helping fertilise the plants – and then the plants eating the animals, as in the end we're all broken down by carnivores and microbes into ever smaller pieces, the building blocks of life, needed by plants themselves.

All energy on earth, pretty much, comes from the sun. Plants and algae have this amazing ability to trap the sun's energy, and through photosynthesis, use sunlight to turn air into sugars. This seemingly simple process captures carbon from the carbon dioxide in the atmosphere and converts it into carbohydrate. In the process, the plants and algae release oxygen. Without plants, animals would have nothing of value to eat. And not enough oxygen to breathe, either.

So, to live, we humans must *always* eat something that has already lived. This might seem like sixth-grade science, but it's important to remember this elemental fact when we look at what we need to keep human populations alive.

Avoiding eating animals altogether, when you look at whole ecosystems, is arbitrary, and not necessarily the best use of farmland. If we *must* eat something that has lived, in order to live ourselves, sometimes, in some farming systems, that may include an animal product.

Meat isn't evil. It isn't murder. It's just what omnivores like us have evolved to eat – and could still eat, if we change the way we look at food and farming.

In this chapter, I'll attempt to get to the bottom of some of the more prevalent myths, and look at what we can do to feed a hungry world.

————·————

According to the Food and Agriculture Organization of the United Nations, 86 per cent of what livestock eat, globally, isn't suitable for human consumption – including crop residues, grasses and fodder crops. This food is something that grows, or is a by-product of something that grows, that we can use to feed humans only *after* it passes through livestock. And if we didn't feed those things to animals, where would those crop residues and the like go? Imagine, for instance, how hard it would be to rid the world of roughly 11 billion tonnes of crop residues produced globally if we didn't feed them to livestock. Eleven billion tonnes is a lot of stalks, stems, chaff and the like, which would take a lot of energy to compost or plough back into the soil, emitting carbon in the process.

Now, beef has been getting a bad rap in the environmental stakes of late. As we saw in Chapter 12, Beef is Killing the World, there are some fairly grim statistics about the water use and greenhouse gas emissions associated with cattle production. Because of this, there's been the suggestion that if we just get rid of meat, we'll fix global warming – an idea I think has been fairly well debunked. As we saw, eliminating all meat eating will only reduce greenhouse emissions by some 10 per cent.

We know that the amount of carbon dioxide in the atmosphere is escalating rapidly. But it's important to remember that all livestock and plants add nothing to the carbon already rotating through the system.

Plants take up carbon dioxide and produce carbohydrate, releasing oxygen, as they've done for countless millennia. The oxygen is inhaled by animals, and some is released as carbon dioxide or methane, or adds to the carbon stored within their bodies. Carbon cycles from plants, to soil, to plants, to animals, to the atmosphere and the ocean. When this cycle is in harmony, carbon dioxide levels are likely to stay relatively constant. New carbon can enter the system, from gas in the ground, from volcanoes, from disruptions to the surface of the earth, but this is only a relatively small amount, and these sources are not constant emitters. (Obviously, the more carbon we can store in plants and soil and the ocean – without acidifying the ocean – rather than releasing it into the atmosphere, the better.)

But there's another source of carbon, which humans have been busily releasing, that has scientists worried: all the carbon that has been stored by plants and algae for millions of years.

During the appropriately named Carboniferous age, around 300 million years ago, the atmosphere was extraordinarily high in carbon dioxide – 1500 parts per million, about four times higher than in the pre-industrial era. With fewer animal species than the present day, this was an era in which huge ferns and vast swamp forests and algae flourished. When trees died, or algae (seaweeds) bloomed and died, they simply fell to the ground, or the bottom of the ocean, and piled up in anaerobic, oxygen-depleted places, similar to the peat bogs we see today. Carbon-rich, these fallen trees and plants and algae, and even some zooplankton in the ocean became buried, and over time, compacted by pressure and heat into thick coal seams, oil deposits and peaty bogs.

Over time, another 40 million years or so, as more and more carbon was taken from the atmosphere as plants respired, the earth's atmosphere became higher in oxygen. And all the carbon that was stored and compressed underground? It is now available to us, in the form of oil, coal and natural gas.

When we burn fossil fuels, we release carbon that hasn't been in the system for a very long time. A time *way* before humans, and modern animals generally.

So all the talk of greenhouse gas emissions from livestock should at least consider the fact that carbon cycles in and out of plants and animals and the atmosphere, as well as cycling through soils and the ocean.

If I feed my cow only grass that hasn't had a fossil fuel–based fertiliser on it, the only carbon it can emit is what was already in circulation. It doesn't *add* to that total.

The *real* problem is our relentless burning of fossil fuels, in the process releasing vast quantities of carbon that has long been locked away, stored underground in fossil fuels. It's this that adds to the total amount of carbon in the cycle.

———•———

I've been reading a bit of scary stuff lately. There's British re-wilding campaigner George Monbiot, warning that if we don't all become vegan we'll ruin the environment forever, and will have the associated population crash to show for it. There are reports on the damage livestock are doing to the world, both with their belching and their eating, with whole forests being torn down to make way for livestock feed. And of course there's the problem with topsoil – our under-appreciation of it, and how we're squandering the soil we do have.

Now, science is only as good as the data, and while I can see there are some massive global issues with food production – meat and dairy included – it's important to look at the whole.

One of the scariest things I've found is research by David Pimentel, professor of ecology and agricultural science at Cornell University in the United States. Pimentel has estimated that if all of humanity ate the way Americans currently eat, with their massive consumption of confectionery and other processed food, the world would consume all known fossil fuel reserves in seven years. That's right, just seven years. Now, the United States is different to Australia, but considering our vast meat consumption (which is about on par with the United States), our own impact can't be understated. And to cope with a growing world population, which seems unavoidable, those of us in well-off, high-meat eating nations need to change the way we eat.

Often studies show the harmful environmental effects of eating meat, particularly beef and other ruminants. I've already dealt with some of the specifics in Chapter 12, but suffice to say statistics on land clearing and methane emissions fall heavily on the side of the ledger that suggests we shouldn't eat any meat, particularly red meat.

I've been reading these studies, and the sensationalist headlines that go with them, for years now. And as I walked our land, and felt the grass grow under my feet, I've had an unease about my meat eating and meat rearing. After all, using simple biological principles, it can take 5–10 kilograms of food to get 1 kilogram of meat from an animal. The higher up the food chain you go, the lower the yield. From the land, we either eat vegetables, or eat things that have eaten vegetables or plants. From our oceans, we may eat a fish that's eaten another fish that's eaten a prawn that's eaten some crab that's eaten some algae. The feed conversion is different for every animal, and generally lower for seafood, because the animals don't need to burn energy to heat their bodies.

If it takes roughly 5–10 kilograms of one food, fed to an animal, to produce 1 kilogram of meat from that animal, shouldn't we just be eating that 5 kilograms of food ourselves, sparing the animal, and producing way more food in the process? Or is there something else going on, where simplistic maths leads to the wrong conclusions?

Having read a bunch of these reports, and their rudimentary starting points, I began to wonder if the science is dealing with the true complexity of the issue. Every farming region isn't the same. Every piece of soil even within our gully has its own characteristics. Not all the microclimates around us are suited to grapes, or cherries, or wheat. If you extrapolate that diversity to all climates, all nations, all communities, is the answer so simple?

And as I walked our land, dug our soil, harvested our crops, I saw the flaw in the science. You simply can't use a whole bunch of land for growing fruit, vegetables or grains that isn't suited to it. It helped me to see this land and then imagine the global implications of such diversity. After all, while our valley and the whole Huon

Valley region in Tasmania was once famed for apples, and they tended to plant every single suitable site to pomme fruit, there was always grazing here. Families owned a milking cow or two, grazed a few head of beef cattle or sheep, in the parts of the country that weren't suited to apples, but suited the animals. The locals were, through poverty and isolation, often quite self-reliant and grew as much of their own food as they could, but animals were always part of that system. One of the five main reasons for owning a piggery in Tasmania in the early 1900s was because you owned an orchard, and pigs could recycle the waste. (Other reasons for running a piggery were owning a dairy, owning an abattoir, or owning a tip, as pigs were used to harness the food content from all kinds of waste products.) And when the bottom dropped out of the apple market, as the fruit trees were grubbed out, the land was used more generally to fatten ruminants. The only money left to make was often from livestock.

So what is this land that can't grow crops? When we bought our farm, most of it had been grazed for decades when there was no money in fruit. There's a tiny remnant orchard, but most of the 70 acres had run livestock on it, including a few pigs at one time. Now, we have a garden, but livestock still dominate the space, as the market garden takes up only about 3 per cent of our land.

There are parts of our farm that grow grass, but you could probably never grow crops. They're overshadowed in winter, are on a steep slope facing south, with a punishing wind that shudders across from the southwest. These parts were never orchard, and for months at a time the moss out-competes the grass. You couldn't grow blueberries on these slopes, despite the shrub's ability to survive in almost swamp-like conditions. You couldn't grow any small fruits commercially. You couldn't plant an orchard, or any perennial crops, in the shade on such a steep slope. You could, in theory, try to sow a summer-only garden, if you terraced the land somewhat, running beds around the contours, though chances are you'd have to dig them by hand. There is absolutely no commercial crop we could plant that would make it worth our time – and that's speaking as an inefficient, machinery-free farmer who makes viable other things often neglected to be grown in this region.

Our cattle, though – well they can range here. They can walk up the slopes of their own accord. The grass, unfertilised and unirrigated, is fed by rain. And these slopes, our paddocks, produce thousands of litres of milk and hundreds of kilos of beef a year with virtually no labour. Our garden, much as I love it, takes nearly two full-time workers, bought-in compost, expensive irrigation and pumps, to produce a not dissimilar amount of food. Food with a lower nutrient density, in terms of its fat and protein content.

So the simplistic notion I often read, that if we just turned all agricultural land over to non-animal crops, doesn't make sense on our farm. We can't grow monoculture crops like soy, or wheat, on our land. We simply can't produce the hundreds of kilos of beans to replace the protein we get from meat. To subsist on this land, 43 degrees south of the equator, in a maritime, temperate climate, you would really struggle to be vegan. But animals, especially grazing animals, can feed themselves on things we can't eat, and produce food we *can* eat, all the while moving themselves around the farm.

It was a pleasant surprise, then, to finally read some research that takes the vast variation of climates and topographies into account when looking at food production. In their 2016 paper published in *Elementa*, 'Carrying Capacity of U.S. Agricultural Land: Ten Diet Scenarios', Christian Peters *et al.* show that in the United States (and potentially globally), on the available land, you can feed more people using an omnivorous diet, compared to a vegan one. Their research shows that the worst we can do in terms of the land's carrying capacity (how many people it can feed) is to have a high-meat diet (no surprises there). But looking at the use of available land, they also found that the best way to feed people a healthy diet is omnivorously, including milk and possibly meat, from animals that *range freely*, rather than living in intensive sheds or feedlots.

Their views reflected what we had instinctually felt on our own farm: that mixed topography needs a mixed farming system to give a mixed diet.

Ruminants in feedlots might be bad for the environment, and not the best use of land – but so is monoculture farming using fossil fuels, especially when often those monocultures (mostly soy, corn and wheat) are then fed to animals that could otherwise eat grass.

———·———

I can't state this clearly enough. The amazing ability of any animal to produce human food from stuff we can't eat, like grass, is something humans are well placed to take advantage of.

Going vegan, with its prescriptive dietary taboos, makes no sense when you look at the ecology of farms, and the best use of available land. It *might* make sense in some cultures, and in some places, but the rest of the globe is better off using a bit of dairy on occasion and using what comes easily from nature, rather than trying to foist some arbitrary animal-free criteria on it.

The research confirms what a lot of farmers already know, but city-based citizen scientists may not understand. Ecosystems exist in harmony when there are plants and animals, and if those animals are able to add nutrient-dense food to a diet, and possibly help keep a balanced farming ecosystem and healthy soil, then so be it. Ignore them at humanity's peril.

Some areas, marginal country mostly, with low rainfall, or high evaporation, or impoverished soil, might allow pastoralism, but would never be suitable for long-term cultivation of grains, cereal crops, vegetables or fruit. These rangelands can support a small number of livestock, which can then be milked, or used for meat. In a world that is fast running out of arable land, and where the usual farming model is to turn fertile land into deserts, we have to be cleverer about the land we do use. As the authors of the 2016 'Carrying Capacity of U.S. Agricultural Land' report noted, 'by definition, grazing land is not arable land', and by this they mean you can graze animals on land not suitable for crops.

Some places are too cold to grow much, and in such climes animals provide energy-dense food. Fattened in the warmer months, and traditionally stored by salting or curing for the winter, animals have long provided sustenance in colder regions.

Quantifying the amount of food produced in an area isn't as simple as measuring it in litres or tonnes. You need to work out not only if an area can grow a certain crop, but also whether that area can grow a variety of crops, enough to meet human needs. A healthy diet is based on having variety, and it also has to contain certain nutrients, including complete protein, complex carbohydrates, and myriad micronutrients. If you look at crops, the type of land, the technology we have to grow them (including seed types and varieties, etc.), it's inordinately complex. So it's interesting to note that when the 2016 study compared the vegan diet with nine other diets (some not well balanced), it required more land use than two omnivorous diets. In other words, a low-meat omnivorous diet that contains possibly some dairy, as well as meat, requires less land than a vegan diet, using current farming methods. Better still were two options for vegetarian diets. (Although, if you rear animals for eggs and milk, it might be wise or at least less wasteful to actually use the resultant veal calves and roosters for meat. Just saying.)

The authors of the 2016 report do note that much of the previous research, when looking at how land could be used, hasn't really acknowledged the practical realities of farming – namely that farmland grows some plant food, and some animal food, all depending on the soil, the rainfall, the temperature, the geography, and the farming methods employed. Some land is best put in rotation, so that you don't deplete nutrients, and one of those rotations, every three to six years, might include grazing by livestock. And the manure from that livestock, along with the types of crops grown, helps to replenish nutrients in the soil.

Now, it's important to note that the study looks at the United States in particular, a nation in which 70–80 per cent of the grain grown is fed to livestock, and the average foodstuff on an American supermarket shelf has travelled about 2400 kilometres (1500 miles) – and at least 10 calories in fossil fuels is expended to produce each single calorie of food.

So what would the results look like if the findings from that report were extrapolated to the world as a whole? To countries with more marginal soils, and where grass (rather than grains) feed more livestock? What kind of diet is the most efficient in terms of land use, and would nourish the largest possible population? Is it really, as the headlines often tout, a vegan diet?

The *Elementa* study is valuable in its recognition that farming systems are complex, that dealing with nature means dealing with animals, even if you're growing grains or tomatoes, and that we can use animals in a thinking, caring way, and still feed ourselves. It is also hopeful, given that about 84 per cent of vegetarians and vegans return to eating meat at some point, which suggests whole nations aren't going to turn vegan in a hurry. It is not, of course, an excuse to continue the enormous overuse of poorly raised meat, or to keep more animals confined to produce ever cheaper, nutritionally poorer meat.

According to the Climate Council, food production is responsible for about 13 per cent of all Australian greenhouse gas emissions. These emissions come from the fossil fuels used to make fertiliser, to fuel tractors and farm machinery, and to transport food, in some instances vast distances to market. Globally, these figures, while harder to pinpoint (because they also include land use changes including forestry), are closer to 20 per cent of emissions. One sad thing is we throw out about 50 per cent of the food we grow, and so squander the fossil fuels used to produce and transport that wasted food. Sydney's Flemington Market spends millions of dollars each year (the true amount remains commercial in confidence) sending once-edible food to the tip – much of which could be used to feed livestock, or make compost, if we so wanted. If global food waste were a country, according to the Climate Council, it would be the third highest emitter of greenhouse gases in the world, after the United States and China.

We grow enough food, internationally, already, to feed one and a half times the world population – so over 11 billion people – by most estimates. The problem with global hunger and malnutrition isn't farming *per se*, but politics. And that's a harder fix.

Long term, if we want a high global population, which in any case appears inevitable, we will have to change the way we farm and the way we eat. In countries such as Australia, we really do need to rein in our currently indiscriminate and wasteful consumption of meat, and move to a more varied lower-meat, lower-emission diet. This will, however, require a reframing of how we view meat, and our entire cultural attachment to it.

And while I think growing plants to feed to animals is a bit of a waste of the earth's resources, I also think growing plants to deep-bury them is an insult to mother nature and the farmers who grew them.

I also, as a farmer and interested consumer, believe there's lots of land that grows animals that can't be used to grow plants for human consumption, and that eating animals (or using their eggs and milk) is a sensible use of the earth's limited farmland.

Finally, I discover, it's not just me, it's also the science.

CHAPTER 17

The No-Meat Diet Isn't the Answer. Neither Is the Attitude

I'm often appalled at the lack of respect shown by meat eaters to those who've chosen to abstain. There can be an insecurity in the responses, particularly online but also in person, to any talk of eating less or no meat, often from those in the farming fraternity. Comments can be spiteful, belittling, dismissive or angry. Rarely is there reasoned debate. Rarely is there respect for the other's point of view.

It cuts both ways. Many vegans take their personal choices and make them not only known, but vociferously so, their food anxieties a major part of their identity. Meat eaters also find attacks on their carnivory akin to attacks on their culture, their way of life. They see an assault on their dietary habits as a slight on their history and their sense of self.

Radicalism risks ruining it for everyone, including the animals in our care.

As I've discussed, I think there are big flaws on both sides of the meat-eating divide. Vegans have blood on their hands from the food they eat, even if that food doesn't include animal products. Meanwhile, at the other extreme, what some people do to animals in intensive farms would be considered callous if not downright cruel if done to a pet.

None of us has an inarguable right to the moral high ground, and so the debate about animal welfare, and about eating meat, is mostly hysterical rather than reasoned and sensible.

I've looked at intensive animal husbandry, its failings, and its proponents, earlier. And here, I'd like to unpick some of the thinking and mindset of those who are against *all* animal use by humans, and suggest that at best it's unhelpful, and at worst, it has done some animals more harm than good. It's not veganism *per se* that is the problem. Rather, it is the way in which some adherents have gone about doing things that has made matters worse.

As a farmer, I like to think I operate on some moral principles, which is why the argument in favour of veganism is endlessly fascinating to me. Vegan sentiments often spring from a genuine desire to minimise animal suffering. But as I've discussed already, simplistic arguments very often belie the truth about what suffers, and what dies, both inside and well beyond the farm gate.

The philosophical debate goes deeper than this. Prominent vegan animal welfare groups also often want *all* animals, be they pets or farmed or wild, to have rights – some suggest equal rights. But in Australia, giving a feral animal such as a fox the freedom to roam, to enjoy natural behaviours, and to source food, increases the suffering of native and introduced animals alike. And if giving all animals rights, the same way we give humans rights, is the ultimate goal, then it ends up being a case of survival of the fittest. Equal rights means that we humans should be able to hunt, outwit, incarcerate and predate other animals, just as other animals do in nature, without individual consequence. Why should humans be the odd one out, in this scenario?

The answer is simple: because we should know better.

Treating different animals differently – even if they're from the same species – is a particularly human trait, one that bioethicists such as Peter Singer have termed 'speciesism'. It is the reason we can treat a pet pig differently to a farm animal, and why we can use some animals for meat, while others we prefer to just admire in the wild. It is confusing. It is contradictory. And yet, it is normal.

Some suggest the vegan argument is so simplistic that it's hardly worth arguing against. But I think looking at the reasons people become vegan helps inform both vegans and those who like to think of themselves as ethical omnivores. If veganism is so good, and yet so few practise it in its pure form, there must be something else going on. It helps to look at food more closely, and some global statistics.

If the world was to turn vegan, it would take generations. In the United States, the number of vegans has risen from about 2 per cent to a still comparatively low 3 per cent, although the percentage of Americans who report as vegetarian has dropped slightly in the past 20 years.

The United Kingdom, on the other hand, has shown a huge surge in veganism, with some reports that adherents have grown fivefold in number in just two years, to 7 per cent of the population, though these data are yet to be confirmed.

Australian data on vegetarianism are sparse and loose. Yes, Google searches for vegan and vegetarian food are on the rise, and vegan processed food is a huge growth area, but there's little solid data about who is fully vegetarian or vegan. Much of the research relies on a phone poll from nearly a decade ago. According to a Newspoll phone survey in 2010:

> *While 5 per cent of people said they were vegetarian, only 2 per cent actually ate a vegetarian diet. This may mean they ate a vegetarian diet most of the time, or that they misunderstood what a vegetarian is.*
>
> *Of the 1 per cent of people who said they were vegan, only one person (out of 1200 surveyed) actually ate a vegan diet. The rest ate animal flesh, milk products or eggs at least some of the time.*

In part, veganism isn't taking over the world because animal products have long been understood to be such a rich source of vital nutrients. Historically, culturally and gastronomically, high-quality protein has always been a critical part of our diet. Protein is essential for human life, and is the dietary element that meat and dairy so readily provide – and in a complete form, as animal-based protein contains all the nine essential amino acids that our bodies cannot make. (You *can* get complete protein from plant-based sources, but you have to combine more than one type to get all the different amino acids you need.)

Meat also happens to be rich in iron. Across the world, 43 per cent of children at five years of age and 38 per cent of pregnant women are anaemic, a result of low iron intake, of which red meat is a ready source. Maternal and neonatal deaths are a major cause of mortality, together causing between 2.5 million and 3.4 million deaths worldwide, in large part because of anaemia.

Globally, 155 million children are stunted and 52 million children suffer wasting, simply as a result of diet, and 88 per cent of countries suffer the burden of several forms of malnutrition. A third of all children in the world are deficient in vitamin A. Of the top 20 dietary sources of vitamin A, there's not a single vegan option. Liver, particularly from beef cattle, is one of the best sources. Our bodies can convert nutrients from plants into vitamin A, but you need to have access to those plants, to cook them, to eat enough of them (because it takes 12 times more carotenoid in a plant to get the same amount of vitamin A from an animal source), and then also consume those vegetables with fat to ensure the nutrients are absorbed into the body. The quickest, simplest way for impoverished people to get enough vitamin A is to give them animal products. Veganism isn't even on the agenda when basic health is at stake.

Meat can't fix all of the world's malnourishment problems, but it can go a damn long way to cutting those numbers drastically. Some of the health problems in wealthy countries are also caused by meat – too much of it, and not enough of other things.

The problem is not with meat in the diet *per se*, but who gets it, and how much of it.

———·———

Along with the nutritional need for high-quality protein and micronutrients, in part the desire for meat has to do with taste – what we as humans have evolved to eat, and how our brains seem to be hardwired to enjoy some of the flavour components and physical characteristics of meat. These, however, should be easy enough to overcome if we truly want a vegan diet. Problem is, most humans *don't* want a vegan diet. Despite veganism being a growth industry in the rich, developed world, it's coming from a very low base; the vegan community's rapid rise is only relative to its previously small size.

I think veganism, while worthy, is on a hiding to nowhere. Yes, you can find studies that say we can reduce greenhouse gas emissions rapidly if we all stopped eating meat. Yes, there's good evidence that most people can lead a healthy life without carnivory. But these studies don't take into account the fact that most people, globally, *want* to eat meat. Most people really enjoy eating meat. Most people, if they could, would feed their children and aged parents meat to help keep them nourished. In reality, becoming vegan is proving to be a cultural problem more than anything. In countries like Australia and the United States, when it comes to cutting down on meat consumption, changing the way we rear the animals for the meat we *do* eat is far more likely to have a greater sway in the long run than telling people to just stop eating it altogether.

Of course, for some, it's easier to be dogmatic. The simplicity of the vegan argument means adherents can forge a kind of religious fervour. This virtue of purpose and groupthink is no different from other people's need to feel part of something bigger and better, something with meaning and on the side of right. Something that could best be described as a saviour complex, where vegans see it as their obligation to save others from their misdeeds. Not all vegans have this notion, probably not even the majority, but the voluble ones do exhibit the traits of people with saviour complex. And in the process they do all the other vegans a disservice.

This religious analogy can go further. Some vegans have a sense of moral superiority, and a tendency to want to convert everyone else to their cause – to the extent that even countenancing some degree of respect for alternative views can seem unthinkable, so that even letting a thinking meat eater consume meat that they believe has come from an ethical system is not an option. The joke, 'How can you tell if someone is a vegan?' and its answer, 'Oh, don't worry, they'll tell you!' comes from a very real place. I've met countless vegans, in real life and online, where it's either their way or you're evil. This blinkered view can be compared to any radicalised religion, where there is no room for disbelief, there is disdain for disbelievers, and there is only the one, true path.

And yet, if you've ever delved properly into the impacts of humans on animals, if you've ever grown food commercially, if you've read much about history, you'll know it's much more complex than that.

As I've said, not all vegans are zealots, and I'd argue probably not most. But the radicalised ones detract from their own cause. On a *Guardian* podcast, I heard Ben Williamson, a senior international media director for PETA – People for the Ethical Treatment of Animals – suggesting that those who get all het up about butter on their vegetable burger may be missing the point. 'The number one thing vegans do wrong is place personal purity over pragmatism,' said Williamson. They can get caught up favouring self-righteousness over realism, when the bread roll itself, or the packaging it came in, or the non-organic vegetables used, may have worse environmental and animal outcomes than the offending spread of butter.

As with any kind of polarisation, the problem with vegan fervour, with sanctimony, is that it ignores the nuances and complexity of the arguments, and can lead to frustration, and even anger, which is vented on anybody who doesn't share the same opinion. And anger, and invalid arguments, aren't going to help win anyone to your point of view.

At Fat Pig Farm, we've been an easy target. I don't know whether the same angry emails we've received have been routinely sent to every chicken and pig farmer in Australia, but I doubt it. People see me on television and want to object to what we do, so they fire off less than polite correspondence with little thought. Some of it is offensive. Much of it is laughable. Most of it is rationally bereft. What saddens me is the lack of intellectual rigour around their arguments, the lack of understanding of how nature operates. I'm shocked at their inability to see the complexity of social constructs, ecosystem harmony and commercial reality.

We've worked with vegans on our farm over the years, two of whom I'm very proud to call good friends. They aren't morally superior, and in fact are honest about their moral opacity (which is a normal human condition, and not solely the preserve of vegans). We've had brilliant discussions about the impact of eating meat, and using animal products. We've tried to navigate logical paths through the complexity of human interaction with the world. Is the cotton in my T-shirt doing more harm than wool in the broader sense? Does wearing plastic shoes have a longer-term effect on more animals than wearing leather? Just how many mice *did* die to grow those lentils, and did the airport you just flew out of cull birds to protect the planes – and if so, does flying mean you're morally responsible for at least some of those deaths?

I'm more than happy to have a discussion with a vegan about their beliefs. I enjoy the insights of those who have thought about these issues and made a substantial commitment with their lives. I am often in awe of their dedication to a cause, when my own will would waver at the first smell of bacon fat or roast duck. I do enjoy pleasures of the flesh. If I did ever give up meat for good, I would miss beyond belief the eye-closing deliciousness of the perfect chargrilled rib eye. I would yearn for the satisfaction I can get from a slice of roast lamb, or the exquisite aroma and texture of slow-roasted pork shoulder, heady with garlic, scented with rosemary, and spiked with the light anise of fennel. I would find it inordinately difficult to forsake pleasure for self-righteousness and gas, as one vegan comedian I saw put it.

So, I get the arguments. I hear those vegans who say the world can't cope with humans, so many humans, eating so much meat.

I really *can* see the point in stopping grain being fed to livestock – the sort of grain that we could otherwise feed to humans – just so we can eat more animal protein. I get that a non-animal based diet has lower greenhouse gas emissions, I really do.

But what I can't understand is why people who have no concept about how food is put on the table are treating me like a murderer, when the things done in *their* name so they can house, clothe and feed themselves could well be more far reaching.

Around us, on our farm and all the farms nearby, the reason there are so many possums and wallabies now is because the neighbours don't kill them. And they used to kill them in their hundreds. The reason vegans don't have to poison rats and mice in the suburbs is probably because the person next door, or the council, or the local eateries and businesses, or someone else – and trust me, it is always someone else – is killing them. If you don't have aphids on your brussels sprouts, someone, somewhere, has probably poisoned them.

Every fig is the result of the death of the fig wasp, which crawls into the internalised flower head and dies. That could be why fig paste, according to the U.S. Food & Drug Administration's *Defect Levels Handbook*, can have up to 13 insect heads per 100 grams. At *Scientific American*, Layla Eplett reckons that 5 per cent of the weight of hops can be aphids, though it's uncertain whether they add any flavouring to beer beyond the hop they call home.

I don't bring this up to be contrary, only to say nobody has the right to the moral high ground; it all just comes down to where your own ethical standpoint lies.

Vegans have animals killed for them. Unless they eat an entirely organic, plant-based diet, where none of the plants have been fertilised by manure or pollinated by bees, then animals have been used or died in the production of their food. Even then, organic farming kills animals, through mechanical means and the use of organically approved chemicals, and animal parts and excreta will inevitably end up in your food.

This isn't to say it isn't worth eating ethically, just that it isn't helpful to preach about purity to others who don't share the exact same coordinates on your moral compass.

———·———

I do have quite a bit of time for vegans. If anything, they do walk gentler on this earth, from what I've learned. They probably are doing their bodies more good than those of us in Australia who eat the average or above average amount of meat each week. They possibly may contribute to less suffering on the planet. Certainly there's less suffering that you can point the finger of blame at. But give me a break if you think the way to convert everyone else to your way of thought is to unleash all your moral fervour on them in an angry fashion.

Attacking people who don't share your views, without first trying to understand their starting point, is neither rational, nor civilised. Being angry, objectionable, humourless and preachy does nothing for a cause. If being vegan is really so good, we will all follow your example. We'll see the light without the threats, without the abuse. But we aren't likely to be convinced by those we see as militant or socially inept. That kind of vegan does the cause no good, only harm.

If you go around labelling every animal product evil, and every animal farming system bad, and make all meat eaters feel threatened because of the choices we make within our culture, our home, our mealtimes, this results in just one thing: we ignore you. We just switch off. Showing horrific, illicit animal welfare footage allows those who shoot it to connect to those already converted, and probably radicalise them even more, and yet in the process alienating the rest of us – the ones who can actually effect more change through the meat we eat. Vegans and vegetarians already don't buy meat, so they have no sway through the capitalist system. It's only those who do buy meat, the thinking omnivores, who can alter the system, through the choices they make, and the power of market demand. The only reason higher-welfare meat such as organic and free-range exist is become some of us buy them. The more we know about the true treatment of livestock, the more we can exercise our power of choice.

It's well known that guilting people into doing things is a slow, painful, often fruitless way to operate. More livestock are confined in more intensive farms now than when animal rights groups first started campaigning. Clearly, the strategy of fear, threats and antagonism has failed.

I have no interest in converting vegans to omnivory, just as I'd hope they have no interest in converting me to veganism. I just want more honest, intellectually rigorous debate about what we eat, how it is produced, and what is affected in the process.

I'm happy for all kinds of people to believe all kinds of things in life, although in my view, simple slogans such as 'Meat is Murder', and viewing large domesticated mammals as somehow intrinsically more 'worthy' in the world ecosystem than insects, mice, snakes and birds, could be described as misguided, or even delusional.

The welfare of animals probably isn't helped with shallow arguments on the value of life from people who probably don't know that when they eat peas, they too, have blood on their hands.

CHAPTER 18

Eat Dairy? You'd Better Eat Veal, Too

In Australia, you can kill a calf with a blow from a hammer ('blunt force') at a point between the eyes and ears – legally according to the RSPCA – and deep-bury or compost the body on your farm.

Why would you want to do that to an otherwise healthy calf? Because it's a boy calf born to pure-bred milking cows, that's why.

Around 400,000 male calves are born on dairy farms every year in Australia, and with little market for them, some farms don't even bother sending them to slaughter. After transportation and kill fees, some farmers only get about $10 per head for a five-day old calf, at certain times in the economic cycle.

Cows – lactating mammals – are just like humans in some ways. They only produce milk after giving birth, and the amount of milk drops off later in lactation. While human mothers have, throughout most of evolution, usually breastfed their young for a period of years, cows tend to wean their young after a few months. In the dairy industry, that means a cow usually gives birth once a year, the calf is removed within a day or two, and the cow is milked for ten months. Heifers, unmated female calves, are fattened on milk or dairy replacer – crazy, but true: it's cheaper to feed calves on powdered milk than the milk their own mothers produce! These heifers will then replace the elder cows that have finished milking. As for the boy calves … well, life looks grim for all of them. Their genetics mean they'll have a poor meat yield once they've grown, compared to beef breeds. And unless there's a market for them to be kept alive, they're killed on farm and composted, or sold into slaughter at about a week old for hamburger or pet mince.

So, vegetarians who eat dairy really should eat veal. If you eat cheese or drink milk, a life has been created to make that possible. In fact, *anybody* who uses dairy should eat more veal. It's such a waste of the earth's resources, such an affront to the animals we rear, to simply kill a newborn or a week-old calf because we have no use for them, and they're simply in the way.

Veal, I know from experience, is controversial. I've received complaints, some not so polite, when I've published veal recipes in the past. Yes, an animal has died. An immature animal, yes, but that's no different to what happens when we rear chickens, ducks, turkeys, lambs and pigs. They are all slaughtered before they reach maturity, because in Australia the market demands relatively low-flavour, tender meat from younger animals.

The complaints are mostly not about the fact that veal is from a sexually immature animal – otherwise I *should* get an angry email every time I write a recipe about chicken, or salmon. No, it's the fact that it's veal. As is often the case, the complainants seem to conflate what happens in some other countries with what happens here.

When I was an apprentice chef, the teachers told us the best, true veal was from an unborn calf. I don't think that was ever a thing, but that was the view back then. And veal was judged by a curiously European standpoint: how pale it was. (So much was this paleness prized that one restaurant I worked in used to crumb pork fillet and call it veal.) In parts of Europe, young calves are still kept in what are called veal crates – isolation pens where they are fed milk, forced to remain anaemic, and not allowed to exercise. This makes for bland, tender meat that has some sweetness from the milk. And it's a system that is thankfully banned in Australia under animal welfare legislation.

Veal here is a different thing altogether. A good veal calf is months old, not days or weeks. It is often called 'rose' veal, because it isn't pale and anaemic, but rather starting to gain a pink hue. Veal can come from either beef or dairy cows, but is more likely to come from animals that aren't going to get as big or muscly as adults – so usually they're dairy animals. Most veal, therefore, is a by-product of the dairy industry, and if you eat dairy, it'd be far less wasteful of lives, of energy, of all our resources, if you ate veal too.

The dairy industry isn't the only one in which unnecessary death comes routinely. The chicken industry is rife with death, but we could, if we organised things differently, at least give some meaning to the lives that are lost. Every male chick from our most popular laying breeds – ISA Brown, Hisex, Hyline and similar – is killed at about a day old. Thrown into mulchers while still alive (called 'quick maceration'), or gassed with carbon dioxide, these budding roosters could one day have made extraordinary stock, or a modest coq au vin. But we dispatch them because meat chickens are so cheap nowadays that it's not worth the bother to rear roosters from egg-laying breeds at all.

And that's not the only waste from laying birds. All those spent hens? The ones that have been laying on farms for a single year, be they in intensive sheds or pastured poultry? They still lay eggs, but not enough to make them commercially viable. So they, too, are sent to the gallows. Some farmers will sell them or give them to home poultry keepers, those who don't have to get an egg from every hen every single day. Some chooks do get sold as boilers, the perfect older bird for better-flavoured stocks, but most birds end up as blood and bone – or possibly salmon food.

Yes, that's right. Salmon, as I related earlier, are often fed chicken fat, chicken feather meal and meat meal in their diet.

When you see how food is produced, there are so many surprising things. Like the fact that it costs more to bottle-feed an orphaned or abandoned lamb than you're likely to make from the meat of the animal at 12 months old. Oftentimes you'll visit farmhouses to find a gaggle of bottle-fed lambs nearby, all bleating with feigned hunger, watching the humans who have become imprinted as their mums. The farmers are looking after them because they feel it's their duty, not because it will make them money.

It takes a sheep about four months of gestation to produce a lamb. It takes a cow about nine months gestation to produce a calf. All the while they are eating grass, exhaling methane and carbon dioxide, using water, and need some human care and attention. Modern farming means chicken eggs are incubated using electricity for 21 days, only to have half of what comes out of the laying hen's incubator automatically discarded because of gender. To me it seems such a waste of our time and efforts to have such callous disregard for the lives we create. It's an insult to the environment that supports not only the animals in our care, but also supports us. I know as farmers we must make choices about lives won and lost. Farmers make decisions to kill rodents, shoot birds, trap possums and cull kangaroos. But to create life, and then treat it with such disdain, appals me possibly even more. To define one bird as worth less than its sister devalues all the birds that have lived. I think it's a sad indictment on us, as a nation and as a species, that we haven't done more to deal with those problems.

It's possible, as we did in times gone by, to breed for dual purpose. For animals that suit not only eggs, but meat. Our obsession with specialisation means that instead of looking for good laying hens, and roosters that can set good meat, we just look at one criterion and treat the roosters as waste. Imagine if we had laying hens that produced 5 per cent fewer eggs (which may mean a proportionate increase in the cost of eggs), but whose brothers were worthy of keeping for the pot? The lower productivity in egg laying could, in theory at least, be offset by money made from meat.

The problem is, everybody specialises, everybody pushes prices down at the cost of the animals in our care. And nobody seems to think that killing about 16 million male chicks a year is a problem.

This could be why I see vegans at least have a clear path to righteousness in some ways. Vegetarians who eat eggs and dairy but who don't eat boiler hens, who don't eat veal, are complicit in this wasteful use of lives as much as the meat eaters in the world. Probably more. Unless we value all the lives we create, it demeans us as farmers, as humans, as animals.

The true cost of meat is something society isn't used to paying. Australia, in particular, has always had cheap meat. For 60,000 or so years the original peoples hunted for what they could find, often with diets high in animal protein, especially in parts of Tasmania. Swan populations in Macquarie Harbour and the suitably titled town I live near, Cygnet, have dropped drastically since white settlement, as their eggs and meat were used for the table. And it's not just Tassie. It has been estimated that by the late nineteenth century, 159,000 ducks were passing through Melbourne markets in a year, once Europeans arrived with their guns and big appetites.

As a nation we then rode on the sheep's back, growing the animals for wool, and gaining the fat lambs that the woolly mothers also provided. Vast swathes of grazing land produced relatively cheap beef. But meat has never been as cheap as it is now, in relative terms, with all the true costs being born elsewhere.

For all the grain that is grown to produce lowest-common-denominator chicken, pork and increasingly beef, the cost may not come home for a while. It'll be there in the loss of topsoil, as monoculture annual cropping is one of the worst culprits for erosion. The cost will be there in the methane emissions from all the poo and waste, only now starting to be used for energy, but sometimes just discarded. It's there in the high fossil fuel inputs for intensive conventional agriculture and intensive livestock production. It's there in our ailing Great Barrier Reef, being hit by runoff of topsoil and fertiliser, as well as through bleaching from climate change. We don't want to pay very much for meat, so male chicks are mulched, calves slaughtered, spent hens discarded, the topsoil blown over to New Zealand.

There are structural issues in food production that we'll have to address as Australia's population – and that of the world – grows, soils lose their fertility, crops lose their nutrient density, and climate change worsens. But I think what shocks me most is the fact that we haven't had the heart to come up with better systems for the animals that we bring to life *now*. Not just the farrowing crates and the feedlots and the battery cages – but farming systems for the animals who don't make it, despite our will to bring them into the world, hoping for girls when we get boys.

With all our nous, with all the money spent on research, it's a sobering thought that we never once seemed to consider ways to reduce the needless wholesale slaughter of unwanted animals by – simply – making them wanted. This might seem hypocritical, because those animals will still be killed later. But I'm interested in minimising harm, in rearing fewer animals for the same gastronomic and nutritional benefit, and spending less time, fewer resources and using our minds to solve problems that are solely of our own making.

We had the will to come up with such complex, intensive systems – farmers and scientists and regulators working together to make cheap food, because that's what society demanded. Surely, if society also demands, the same alliance can fix the problems we've created in the first place.

And in the meantime we can all help by eating veal.

CHAPTER 19

Fake Meat

If the whole world wanted to eat like we do in Australia, with our love affair with meat, along with our hunger for other high-input food, there would simply not be enough farmland available. In other words, we consume way more of the earth's resources, per head, than we should. We'd need about 1.6 planet Earths to grow enough food to feed the average Australian diet to the global population. And much of this has to do with our insatiable appetite for meat. While animals take up about 80 per cent of the world's agricultural land, they only produce about 20 per cent of the world's calories, or energy supply.

It's understandable, then, that there's a lot of research being done on meat alternatives, and meat grown in labs. There's no shortage of money flowing in, much of it from Silicon Valley in the United States, a place over-represented not only in the top tax bracket, but also as investors in the vegan and vegetarian industries. The aim of various companies is to either produce meat without having to rear and slaughter an animal, or to make meat-like products using plants, such as the Impossible Burger, which 'bleeds' when you cut or bite into it.

Lab-based meat has been made possible by vast advances in stem cell research and the like over the past couple of decades. It's been dubbed 'test tube meat', and by opponents as 'Frankenmeat', and uses the principle that if you put muscle cells in a lab, feed them and get them to divide, you'll end up with meat. Great in theory, but pretty hard to do in practice. Real meat, from an animal, is incredibly complex in structure, with plenty of collagen, fats, peptides, blood vessels and more going on. All of these can, in some way, contribute to flavour, and very much so to texture. It would be unlikely that science will unravel the mysteries of how meat is constructed in an animal and be able to replicate it in a way that is as efficient as the animal that the method was copied off. We can't even intensively farm meat animals today and keep the same qualities in the end product as free-ranged animals: according to a report in *The Guardian* in June 2005, scientists from London Metropolitan University have found that a chicken from 2004 contained a third less protein than a chicken from 1940; interestingly, the modern chicken also had a third more calories.

Trying to copy complex systems in a lab, then commercialise them in a factory, is incredibly ambitious, and I don't think we'll be seeing whole cuts of lab-grown meat, like steaks or chops, complete with all their natural intrinsic qualities, any time soon. But perhaps I'm wrong. There's more than one company trying.

Mostly, though, the work on cultured meat, lab meat, is about producing meat of the lowest common denominator – mince, or ground meat as it's known in the United States and increasingly around the world. The meat they're trying to replace is the commodity mince used in burgers and, less so, 'surprise bags' (sausages). It's the meat churned out by feedlots around the globe, and also the fate of every clapped-out dairy cow. If there's any meat to replace, then this strange burger patty meat might as well be it. The environmental costs of this meat are currently very high, the eating quality is low, the texture of the meat is already compromised by mincing, and both flavour and texture are usually hidden in a bun with other ingredients such as pickles, sauces and spice.

The burger trade is huge business, however. There are at least ten companies in the global race to produce the first commercially viable lab meat, dubbed 'clean meat' by some clever marketing crew. Mostly they aim at reproducing muscle cells, then adding flavourings and fat to give the meat some of the 750–1000 flavour compounds that it would otherwise derive from nature. Some tastings show we can't tell one species of animal's meat from another in the absence of fat, and it's the fat that defines our major perception of meat taste. Lamb and beef, without fat, taste remarkably similar.

Making things palatable is the realm of food technologists, drawing on a cache of chemicals. But the technology for replicating muscle fibres already exists; making it affordable is the great challenge. In 2013, the world's first lab-grown burger patty was produced – at a cost of around $400,000. Perfected by Dutch scientist Mark Post, the burger patty made headlines around the world not just for its staggering price tag, but for the breathtaking reality that meat could be grown outside of a body.

More recently, another lab-meat company, Memphis Meats (funded at least in part by Bill and Melinda Gates and Richard Branson), which is working on making duck and chicken meat from stem cells, reported that it now costs only $6000 to make 1 kilogram of meat, with costs falling every few months.

But these stem cells – and the issue of how to feed them – remain an ethical problem for many vegans. Currently, to create cultured beef or so-called 'clean meat', an animal still has to die. That animal is a preterm calf, which has its foetal bovine serum removed to help feed the stem cells. The calf can't survive the process. There are alternatives, but for now they're very expensive, and not as good at the job. These are problems that science will probably solve, if people want lab-grown meat; certainly there's strong motivation to be the first to do it.

I reckon there will be a place for these products, just as there's a place for frozen pizza and tinned spaghetti. But I am cynical. For all their money, and the expensive food photography and styling, the homepage for Memphis Meats says it all. Shot with a short depth of field and some backlighting, their main photo shows a plate of food, with a couple of what appear to be green olives, some strands of tagliolini, a few cubes of diced capsicum, and a gluey, greyish brown, indigestible-looking blob of their product on top. No amount of photographic wizardry makes it look good. This is lab grown, and more to the point, it's lab manipulated, with flavours and colours added to make it palatable. It might replace some kind of processed meat, but it's not meat. And the term 'clean' implies traditional meat production is somehow 'dirty' – which might work in a marketing sense, but denies millions of years of human evolution. All those chemical compounds and micronutrients that you naturally receive in the meat from an animal are exactly what our digestive system is designed for.

Even the website for Aleph Farms, a cultured meat company from Israel working on manufacturing steaks from four types of meat cells, rather than just one, have a graphic of a farm barn and free-range livestock on their homepage. (Aleph Farms, by the way, are backed by the Strauss Group, which has an association with PepsiCo, and make Doritos and Cheetos.)

When I think of scientists trying to make meat, I think of the

Cloaca Professional (2010), an installation by artist Wim Delvoye at Hobart's world-class art gallery, MONA. It consists of a large machine, stationed in a darkened room, with parts that whir and ferment and digest food, and 24 hours later out comes a turd, something resembling faecal matter. It's interesting because this entire machine, affectionately called the 'poo machine' by locals, is still simply an approximation of digestion, and produces no useful energy, and only burns energy, unlike the digestive tract of a mammal. Think of the miracle of milk: how a cow turns cellulose from grass, which we can't digest, into a high-quality, highly digestible substance, milk, in a period of hours – something science has failed utterly to replicate. The cow does this while using energy from food to not only digest the food, but also heat its body and run its other functions as well. The Cloaca Professional burns energy just to digest.

I'm not anti-science, just sceptical. I think the most efficient way to grow meat is probably on an animal. Funnily enough, a 2017 study found that it was more energy efficient to produce chicken on a chicken than fermenting it in lab.

Growing nutrient-rich food in a way that doesn't ruin the world, or our health, is our challenge. Cultured, lab-grown meats may be one way to wean people off their addiction to real meat, to cut down on the number of animals eaten. Yet lab meats allow our eating culture to continue unabashed, placing little emphasis on provenance, on freshness, on nutrient density, and more emphasis on convenience, without conscious recognition of the impacts of our eating culture on the world and the greater environment. These products may well be a good stepping-stone to a more sustainable existence. I don't think, though, that I'd consider such highly processed things real food.

Parallelling the rise of cultured meats is the increased research into plant-based products that are manufactured to look like meat.

A long way from the lentil patties I used to make at uni, these are funded by start-ups from Europe to the Americas, including that home of gastronomy, Silicon Valley – with the simple stated aim of preventing animals being harmed to produce food (even though, as we've already seen, grain and pea growing also do cause animal deaths). There's a whole lot of money being spent on this end of the game, too, and in some circles, these animal-free products are already replacing the meat that was once being eaten.

I'm not sure if the amount spent on this research is the best use of resources, but if anyone can improve on the vegan sausages and 'facon' (fake bacon) that I've had the displeasure of trying over the years, then perhaps it's not a bad thing. More to the point, however, is that these companies are trying to work around our cultural attachment to meat, while weaning us off the animals themselves.

Beyond Meat makes burger patties by passing ingredients such as pea protein, coconut oil, canola oil and water through a heating and cooling system that their CEO Ethan Brown compares to a cow's four stomachs. If you've ever seen what comes from a cow's four stomachs (and I regularly step in such a product as I pace the farm), then I'm not sure that's a selling point.

It's interesting, but presumptuous and telling, that Brown compares an industrial system to the majestic simplicity of an animal's gut. Essentially they've taken what an animal just does, and are trying to replicate it in the lab. Maybe it's a noble aim. Maybe it's going to decrease animal suffering. I think it'll just burn more fossil fuels and produce something that provides protein and energy for our bodies – a soulless product that provides little joy, and adds nothing to our cultural fabric. Maybe, just maybe, it's the new way.

Impossible Foods is another high-profile brand of plant-based meat products, most famous for their burger patty that, when cut, weeps a reddish substance that is reminiscent of blood. Impossible

Burgers are made from a bunch of things, including gluten, potato protein, soy sauce and heme, a flavour compound made using soy and a genetically engineered yeast. There's a great photo of stainless steel tanks and people in lab coats and space suits on their website, which is appropriate given that the products offered by both Impossible Foods and Beyond Meat are heavily processed foods, involving fossil fuels to grow and transport the ingredients, and to process them at the factory. They may be better than some beef in some respects, but here's a list of ingredients in the Beyond Meat burgers:

> *Water, Pea Protein Isolate, Expeller-Pressed Canola Oil, Refined Coconut Oil. Contains 2% or less of the following: Cellulose from Bamboo, Methylcellulose, Potato Starch, Natural Flavor, Maltodextrin, Yeast Extract, Salt, Sunflower Oil, Vegetable Glycerin, Dried Yeast, Gum Arabic, Citrus Extract (to protect quality), Ascorbic Acid (to maintain color), Beet Juice Extract (for color), Acetic Acid, Succinic Acid, Modified Food Starch, Annatto (for color).*

What was it that Michael Pollan, the famed American author of *In Defense of Food: An Eater's Manifesto*, has said? 'Don't eat anything your grandmother wouldn't recognize as food.' (He also advised, 'Eat food. Not too much. Mostly plants.' I'm a fan.)

Yes, I am sceptical of products made by these large companies, because food manufacturers usually don't have anyone's best interests in mind except their own. But to satisfy the fast-food market – the market that currently cares little or not at all about the origins of their meat – I'm all for products like Beyond Meat filling a niche. They're allowing people to eat things they find culturally acceptable, without having to eat meat. They're reducing the number of animals needed to be raised in the world – undoubtedly a good thing – and possibly cutting down on some of the environmental damage being done by the beef industry, too. I do, however, find it telling that the Beyond Meat website has more than one comment from purchasers saying that they drove over an hour to buy their burger patties. It seems some environmental costs are all a matter of where you stand (or park) at the time.

———

Mark Post's original beef burger, the $400,000 one, is a sign of the times. It's a phenomenon foreseen by none other than former British prime minister Winston Churchill, who wrote in 1931, for a selection of predictive essays called *50 Years Hence*:

> *We shall escape the absurdity of growing a whole*
> *chicken in order to eat the breast or wing, by growing*
> *these parts separately under a suitable medium.*

So this whole animal-free meat thingo has been on the cards for a long while now, albeit materialising a few decades later than Winnie first speculated. Post's original lab-grown burger, despite being described as dry and anaemic, showed what can be done, if we put our minds to it. Whether it's worth putting so much effort into growing something that already grows – something we have over-exploited, ignoring nature's warnings in the process – remains to be seen. Consumers, when surveyed about whether they would eat fake meat products, express concern about cost, flavour and the unnatural nature of the product.

I'd rather we eat less meat. Better meat. Good-flavoured meat that satisfies our cultural leanings and our physical yearnings. Meat that doesn't drag more long-stored carbon out of the earth and put it into play. Meat that is something joyous to celebrate – not merely devour because it's the next best or the newest or the one with the cool marketing.

'If you make food that tastes really good, you win,' *The Guardian* quotes Josh Tetrick, CEO from JUST, a maker of vegan 'egg' products such as egg-free mayonnaise and egg-free scrambled eggs, as saying. Tetrick has been called the 'Elon Musk of condiments', but it sounds just like any other corporation making processed food to me. Perhaps because I was raised on food that tasted like food, and on our farm we grow and cook using fresh ingredients, I reckon food that doesn't try to be something it's not can taste good, whether it contains animal products or not.

I have to say I'm more than a little suspicious of the motivations of big companies, because they got us hooked on processed food, they've trained our palates to love sugar and salt more than ever before, and globally they have the clout to do what a carrot, a pat of butter and a mango can't do: namely, sell us more stuff. Processed food of any kind just removes us even more from the origins of our meals.

I'm also suspicious because this kind of science has traditionally added precious little to our gastronomic trajectory. For all the money, effort, time, science and research that the best minds in food technology have put into making a butter substitute, the only thing they've come up with is margarine – something that's not even a close approximation of the real thing, an ingredient offensive to those who like real flavour. I like science. I like scientists. I just don't feel comfortable letting them in the kitchen.

All of these companies, all of this money, all of this processed food, however, serves a purpose. And that purpose may well be to replace more traditionally made – dare I say more natural – food. It isn't, however, going to have much of an impact on the world as a whole, because the world as a whole isn't eating processed food grown in monocultures and sold by conglomerates.

It's important to remember that even as I write, 70 per cent of the world's food is grown on small farms, and most of it by women. Most of the food in the world hasn't passed through a factory. Most people don't have access to the kinds of packaged, handled, chemist-driven food products that grace the shelves in Australia. Many people in developing nations have issues getting enough food, and certainly enough of the *right* kinds of food, without Silicon Valley start-ups trying to out-meat meat.

Globally, there's little choice but to reduce our meat consumption. I think that'll take a change in mindset, and a change in our day-to-day practices in huge meat-eating places like Australia. Sure, manufacturers need to change too, but a bad burger made from an intensively farmed cow is hardly going to be missed when replaced with a bad burger made from pea starch and bamboo extract. Neither of these can be expected to add to our cultural identity, our integrity, our sense of self. The new burgers promise to fill our tummies, and empty our wallets, and maybe lessen the damage done to the environment at the same time. But they're not the magic bullet that will put good food on plates around the country.

That change has to happen in the farms, shops, butchers and restaurants of our nation, and ultimately change has to happen in the domestic kitchen.

CHAPTER 20

When the Hunter Becomes the Hunted

Something happened when I got my gun licence. A whole secret society opened up. I'd already suspected this society existed – not in a *Da Vinci Code* kind of way, but because I'd been hanging around those who already recognised the signs. A camouflage hat in the back seat of a ute. A mounting for a gun sight on the kitchen bench. A spent shell at the opening of a gate. My mate Ross, a chef and hunter, would see these things on our travels and strike up conversations with the owners. About deer. About wallaby. About things to do with guns and rifle ranges. And more specifically, about hunting.

The reason this society is so secret is because in many parts of the country, for the urban majority, guns are a topic of division. With fairly stringent gun laws, Australia has managed to curtail gun violence, but many people argue that they can't do their job properly if the laws are too strict. And by many, I mean farmers. Faced with mobs of roos, or wild pigs – or worse, packs of feral domesticated dogs – there are few options for control. Poison and fencing aren't as effective as guns, which are used to kill those feral animals that make farming more difficult, or to kill native animals, which compete with domesticated animals for food. You could simply argue, and many do, that we should just get rid of the domesticated animals and let the native animals have the grasses. But roos and wallabies and possums eat crops – grain, fruit, vegetables – too. If you eat food, often the farmer has had to kill an animal to protect their crop, and oftentimes that's done with a gun.

Now, I'm not going to argue whether a farmer needs a semi-automatic weapon or not. Gun control advocates rightly point out it's these fast-action guns that cause the most carnage in mass shootings in the United States. What I will discuss here is the use for a gun, and who will wield that gun. On our farm, while we can use our gun for crop protection, the main purpose of my rifle is to be able to end the suffering of one of our animals if it is sick. For too long I called on others, kindly neighbours and friends, to do this work for me. But the lives of domestic animals on my farm are my responsibility, as are their deaths, and making these timely and reducing pain and suffering go along with that obligation.

There's a certain attitude amongst those who don't own guns, and don't know (or don't realise they know) someone who has a gun, that all gun owners love to kill. And from the outside, looking in at the hunting fraternity, there's a feeling that there are a lot of people only in the game because of blood lust.

Now, much as I, and perhaps you, may disagree with these people, what they do when they track, stalk, scope and shoot wildlife is no different, really, than Australia's favourite pastime – fishing, something about 15–20 per cent of Australians do every year. The only difference is the animal at the centre of the chase, and our attachment or otherwise to it. There are plenty of people who will happily catch a striped tuna, and yet they see the killers of fallow deer (who can forget *Bambi*?) as barbarians. I've a friend who is vegetarian, for ethical reasons, who loves nothing much more than to thread a hook with bait and bring in a flathead, though it's not her that eats it.

The overarching problem with guns, with killing, with hunting, in my view, is with perception, and with how that translates into action. Duck hunters are chased by animal activists from the first day of the season. Kangaroo cullers are demonised by international organisations. Those who shoot deer are often portrayed as savages. And no doubt, there are those amongst the hunting community who get a kick out of killing for killing's sake. And I really, really don't understand that.

But let's look at it from the animals' point of view. Take ducks as the example. Duck hunters are demonised, the subject of annual protests, and chased through wetlands at the start of every season. But the owners of millions of farmed ducks housed in sheds in Australia are somehow not subject to anywhere near the same public scrutiny.

Wild ducks at least get to express their instincts. Shedded ducks don't. Wild ducks contribute to the ecosystem during their lifetime. Shedded ducks don't. (If anything, they compete with, and possibly damage, native ecosystems, given every farm was once a native ecosystem.) A duck in the wild suffers less at human hands than a duck in a shed. Wild ducks have a stronger, bolder flavour, meaning that when you cook them, you need less of the meat to feel satisfied. And a duck in the wild, just like a shedded duck, will die at some point. With the wild bird, it's only the human-caused death that animal welfare groups object to (and if the hunters don't eat the birds, it's a shameful act from pretty much every point of view).

Some argue that it's only younger, less able ducks that get shot in the wild, and they might not have survived to adulthood anyway. A spurious argument, perhaps, but one that does have some science to back it up.

Now, I'm not going to argue the intricacies of ecological damage from duck shooting here. I'm not sure I ever support duck shooting in wetlands, and never using lead shot, because of its poisonous effects. But I will argue the duck's life is better in the wild than ducks reared intensively in sheds.

By contrast, let's take a look at Australia's two largest producers of duck meat. First, there's this, from the Luv-a-Duck website:

> *Our ducks are housed in large, enclosure like barns that allow natural light and fresh air to flow in and also provides room to move around and for the ducks to socialise.*

Read, they're in sheds and can move, so they're not in cages.

> *Trained stock-men take care of the duck's welfare ensuring that they are comfortable, relaxed and that there is always dry and clean litter to walk on.*

Translation: staff ensure basic welfare.

> *By growing the duck inside they are protected from adverse weather conditions and predators that include foxes, feral cats, crows and magpies. They are also protected from disease transmission from migratory birds which is a serious issue.*

So this suits the farmer – but given a choice, would the ducks live in this environment? We wouldn't know unless Luv-a-Duck opened the door to let the ducks decide. And from the looks of it, Luv-a-Duck don't.

The Australian Competition and Consumer Commission (ACCC) didn't like the way Luv-a-Duck was promoting its products, with the company implying that their shedded ducks were in some way like wild ducks in their ability to roam free. In 2013, the ACCC fined Luv-a-Duck $360,000 for saying their ducks were 'range reared and grain fed' in the 'spacious Victorian Wimmera Wheatlands'.

Around the same time, Pepe's Ducks, another intensive operator, and Australia's largest duck producer, was fined $400,000 by the ACCC for similarly deceptive conduct, claiming their ducks were 'open range' and 'grown nature's way', when they clearly weren't. By comparison, wild ducks actually are 'grown nature's way', and they *do* get access to open water – a freedom usually withheld from intensively farmed ducks.

The mob that secretly film atrocities inside intensive farming sheds are the same ones that fight tooth and nail to stop the duck hunt as well. I'm probably sympathetic to their notions, but the messaging that all animal farming is bad, and that all hunting of naturally reared animals is bad, won't stop people eating meat. Condemning factory farming for failing to give animals a natural life, then condemning those who harvest animals that *have* lived a natural life, is contradictory.

What does this mean for the average consumer? All it does is push animal welfare into the 'too hard' basket part of the omnivore's brain, so they ignore it. All it does is make intensive farms even more secretive, and even less likely to open the doors, because they think everybody interested in welfare is a vegan. And it makes it much harder for the conscious omnivore to make ethical choices, or to even see what happens on their behalf.

———·———

Where once we were all hunters, now it's only the few. I know that I find little joy in hunting, but I also think that killing animals to protect crops, or manage wild lands, and then wasting their meat is an abomination that insults mother nature.

Perhaps it's the change in emphasis, the change in need that has altered our views. Today there's the idea that you hunt so you can kill, rather than you have to kill so you need to go on a hunt. The only meat available to our first peoples for 60,000 years was from hunting, and game meat sustained our early European settlers, too. Only recently has it become publicly unacceptable to say you shoot to eat.

There's a strong argument in favour of eating the animals that are suited to our climate. Kangaroos, with their soft impact on the land, don't destroy the lichens that hold fragile desert sands together, and so are less likely to damage soil structure. They can breed up quickly in times of plenty. They can forage more widely, and thrive in the mix of scrub and grass that dominates our marginal lands. Maybe we should eat more of them, and fewer cows? The issue comes, as usual, with quantity. There's nowhere near enough kangaroos to replace our beef herd, and a lot of people have issues with eating the animal that is on our national coat of arms.

———·———

While admitting to hunting is socially awkward in urban settings, I would argue that ethical hunting – where the meat is used, the method is appropriate and the numbers of native animals taken strictly controlled – deserves a better rap.

Let's face it, each year in Australia we cull about 20–30,000 camels in our arid areas, and most are just left to rot. Same with most of the 5 million or so kangaroos that are killed each year to protect crops and grasslands. In 2017, in Kakadu National Park, authorities culled 6000 feral horses, donkeys, buffalo and pigs to better protect the iconic area. And there are more wild pigs, by some estimates, than there are humans in Australia. In 2012–2013, marksmen shot 30,000 feral pigs from helicopters in Queensland alone, the bodies left to rot. Is this utter waste of life really a good and sensible outcome? There is an export market, for feral pigs at least, with one Queensland operation, Game Meat Processors, exporting 1000 pig carcasses a week to France and Germany, where much of it gets turned into smallgoods. In 2015 the owner reckoned he could sell 5000 a week, if he had the hunters on the ground to get them, but a lack of accredited hunters was hurting his business. And still feral pigs are a problem for farmers and for wildlife.

There will be those who argue that we, as an intelligent, organised, successful species, should move beyond hunting. That we should outgrow our need to chase down and kill wild animals simply for the table. That we should at least farm animals rather than shooting something as noble as a wild beast, and possibly even desist from using any animal for food. And you know what I think of that idea, the meat-free diet, from other chapters of this book – that removing animals from our culinary repertoire is arbitrary and meaningless within the bigger picture.

Hunting may allow us to utilise Australia's land mass better. It may help prevent the damage that cloven-hoofed animals do to some of our fragile soils. It may be a much smarter, more efficient use of a resource, both the native kangaroos culled, and the feral animals culled – animals that at the moment are simply left to die, or are not valued enough for the lives that are taken. On King Island, with a population of just 1600, they kill over 80,000 wallabies in a single year, most just left to rot. Tasmanian farmers kill about 1.3 million wallabies every year, and little of it ends up in human mouths. As on King Island, most of them just go to waste.

Hunting isn't all bad. It isn't always done by people who can't control their blood lust, or by people you wouldn't want to break bread with. In terms of animal welfare, it's certainly arguable that hunters eat the most ethical meat – animals reared as nature intended, animals that were eating a complex diet, and living within the constraints of the environment as their instincts dictate.

Hunting can seem barbaric, especially to those who've never grown food commercially. But I know of vignerons, orchardists, vegie gardeners, graziers and pig farmers who all have a gun and kill creatures that invade their space. As we've seen, animals die all the time so we can produce food. But hunting for trophies, hunting for mere sport, hunting to leave carcasses to rot and go to waste is an affront to the land that carried those lives. If we can make sensible use of these resources, we'd be mad not to.

Killing wild animals may, in some circumstances, be a valid way to get high-quality protein. It will take a cultural shift for that sensible use to become reality; to understand that not all hunting, like not all farming, has the same ethical starting point. Hunting could, with no increase in numbers currently culled, add so much to our dining tables if we learned to utilise the animals currently discarded. But unchecked, and without valuing the work done to protect crops and rangelands, hunting could well be written off as just another bad consequence of our insatiable appetite for meat.

CHAPTER 21

It Was a Former Vegan Who Taught Me How to Kill

In my struggle with the ethics of what we eat, in particular meat, I've found that I'm not alone. You could fill a small library with all the scribblings over the millennia on the philosophy of eating. And in the attempt to work out my own personal food choices, I've discovered that I often have more in common with those who abstain from meat, than many who consume it.

For a long time my food choices were driven by quantity, my appetite never assuaged by an ordinary-sized meal. Later, as a chef and food critic, I became obsessed less with quantity than quality. And now, after ten years on the land, I find that my musings on food, the pleasures I gain from a meal, are also driven by my understanding of farming and the impacts on the land.

This personal journey possibly mirrors a wider global attitude to food. Poorer people are, by necessity, driven first and foremost by the desire to eat enough food. And then enough of the right food. Once we have food security, as individuals, as societies, as nations, then we can focus on the joy it brings; the cultural aspect of eating and the desire to bring pleasure as we cook and serve edible gifts crafted by our hands. And finally, when we have the time, the wealth and the surfeit, we can think about the ethics of *what* we eat. As I guess we must, as intelligent social beings who – most of us anyway – want to think that what we do to fuel our bodies, to bind our culture, won't cause unnecessary harm.

Strangely enough, my journey into the world of meat has driven me closer to those I wasn't expecting. In my circle, people who care enough about the origins of their meat, I have befriended vegans and ex-vegans. People who kill now, but who once tried to eat no animal at all. An obsession, perhaps at times an unhealthy obsession with food, has led those interested in the origins of their meat to either abstain altogether, or to take matters into their own hands. It was a vegan, or rather, an ex-vegan, who taught me how to lop the heads off chooks, eviscerate them, and dress them for the table. In fact, it was two ex-vegans who bloodied me with the task of decapitating a dozen birds destined to be dinner.

Caring about what you eat can lead to various strategies. These days, I think I would be hard-pressed to eat much meat if I had no idea of where it came from. I'd be too worried about the lives impacted by consumption of the wrong meat, be they the lives of the farmers, the communities or the livestock. I also don't think I have the moral fortitude to become vegan – and I don't think it's necessary, as I've outlined earlier, to save the world – but I do think I'd moderate my meat eating even more if the farming of animals remained hidden.

Caring about what I eat is something I've had some luxury to enjoy for many years now. But I was still surprised how ethical eating and veganism are more closely aligned than one might think.

Take Tammi Jonas, for example. This Victorian pig farmer and butcher lived the vegetarian life for six years, and is now active in the Australian Food Sovereignty Alliance and organiser of the annual Slow Meat conference. She's built her own on-farm cutting shop to transform her free-range pigs into pork, not – at first glance – the most logical step for someone who previously abstained from meat. There's cheesemaker Nick Haddow, who was vegetarian when we first met two decades ago, and who now has a stake in a butcher's shop in Hobart. Take Grant Hilliard and Laura Dalrymple, who run the most ethical butcher I've found on my travels, Feather and Bone in Sydney. These are butchers who are famed not only for the incomparable sourcing of sustainably produced meat and their deft butchery of it, but also because they recommend you eat less meat. This goes against every message from those in the peak bodies for all our animal industries, the Australian Chicken Meat Federation, Australian Pork Limited, and Meat & Livestock Australia. The overarching brief of the peak bodies is to get you to eat more meat, and specifically more of *their* meat.

The total turnaround experienced by former vegans who now kill their dinner might seem contradictory, but there's no ultimate right or wrong when it comes to making your own food choices. I'm constantly conflicted by what is acceptable when it comes to ethically produced food. The most fraught example is possibly the most infamous: foie gras, the fattened liver from ducks and geese, loved by connoisseurs and despised by animal activists. The process involved in producing foie gras (meaning 'fatty liver') isn't allowed in Australia, but is still common all over the south of France, and the products are available in Australia.

The practice used to make foie gras is called, tellingly, *le gavage*, or 'force feeding'. The birds – mostly ducks these days, because they are easier to handle and grow quicker than the more traditional geese – are fed grain, then force-fed grain for the last part of their life, to engorge the liver. A funnel is inserted into their mouth, and the grain poured down so the birds swallow an inordinately large amount of it – so much so they have trouble walking after a feed; then they waddle off to digest it. The liver, struggling to digest so much food at one time, becomes engorged and fatty, and up to ten times as large as a normal liver.

Now, the reason foie gras exists, many would argue, is that hunters first noticed the engorged liver of wild ducks that were caught just before they migrated for the winter. Having stuffed themselves silly, the ducks would be fuelled up for the journey ahead, not dependent on feed stops all along the way – their livers enlarged, and delicious. Farmers saw this, enjoyed the livers, and tried to emulate the process. So far, so good.

I've actually been to a few duck farms that make foie gras in France, and watched *le gavage*. On these farms, the ducks ranged mostly free, though they were handled regularly to train them to be unafraid of humans. In the last two weeks they were fed the grain, manually, by the farmers, through a funnel. The birds literally ran up to the farmers to be fed, at least at the start of the period of *le gavage* when they could still run. Later in the process they waddled enthusiastically. And while the engorged liver is classified as 'diseased' in medical terms, if the feed level goes back to normal, the liver returns to its usual size and fat level, and the bird regains its full health.

So is foie gras ethical? The birds I saw seemed to relish their force-feeding, clamouring to get fed. This gorging comes from a natural instinct, the pre-migratory fattening, albeit probably taken to an extreme. And the bird isn't so sick that it will die – it is just struggling to digest so much food, and hence the liver becomes fatty.

Now, there's a whole bunch of farms in France that don't look like the ethical free-range farms I've been to. Many use robotic feeders and shedded ducks to make foie gras for the mass market. And I'm not totally convinced that a diseased liver should be the aim of any farmer these days (though France has very different views on animal welfare – except, perhaps, for their dogs). But there's also a Spanish farmer, Eduardo Sousa, who makes foie gras by simply giving his geese a home that is too enticing to leave, and food that is too good to not eat.

Eduardo Sousa bases his style of fattening on an old local tradition that – he reckons – has been used to produce foie gras, on and off, for more than 500 years. He's recreated a thriving home for geese, and his foie gras – *quelle horreur!* – has also won a French award for the best out there, the acclaimed Coup de Coeur. And none of his geese are force-fed; they simply just gorge themselves on what's there and what grows in his paddocks. They go through the normal pre-winter fattening and liver enlargement that nature has inscribed in their DNA, and he kills the geese just as the season is about to turn, just before the geese flee to warmer climes.

Is this different from a feedlot where the cattle will gorge themselves on grain? I reckon so. Is it better than robotic feeding for ducks that don't see daylight? Probably. Should I eat foie gras, at all, or only when I know the producer? Um, maybe. Probably. Um, possibly.

My issues with foie gras, while more academic than practical because on our own farm we pretty much just eat what we grow these days, are very much First World problems. But so are all the ethics of eating meat. Developing nations, generally, have a lower environmental footprint. They might be struggling to get enough food, or the right food, and to them the luxury of ethical eating might be as strange as growing food on Mars. But my grappling with foie gras, a single, unnecessary ingredient that doesn't even come from around here, does show the personal nature of choices that can have big societal and global impacts. What we eat, how it's raised or farmed, and what we do with it once we've got it, are micro decisions that we're part of every day. They have impacts – small, but obvious impacts – on much larger macro decisions, such as how farmers farm, what they farm, and what is acceptable to do to the animals, and to the environment, and to the society that farms and consumes them.

———·———

Caring about what you eat is a luxury many can't afford. But for those who can, it's worth noting that big change is possible from unlikely quarters, where you might least expect it.

One of those quarters is an American woman named Temple Grandin, the world's most famous abattoir designer. Grandin, who is autistic, was virtually non-communicative as a child, unable to speak until she was three and a half. She discovered, though, that she had an affinity for livestock on a visit to a family ranch in Arizona as a kid. She also came across the squeeze chute, a device for restraining beef cattle by pressing on their flanks so they can have their health checked, and be treated if necessary. Having felt her own pleasure in confined spaces, and stress in human hugs and wide spaces, Grandin noticed some animals in the squeeze chute relaxed once a gentle pressure was applied to their sides. She gave it a try, and discovered it lowered her own stress. A device that pressed against the flanks helped decrease her worries, too, so she invented her own 'hug machine' based on similar principles. It was from about that time that she started to realise she had similar thought processes to the animals in farmers' care. (The BBC even filmed a documentary about her, entitled *The Woman Who Thinks Like a Cow*.) Grandin is now an Associate Professor of Animal Science.

Dr Grandin believes it's her different way of thinking, her autism, that allows her to sense the world more like an animal, with simpler emotional responses, and a state of anxiousness not found in the general population. At first glance Grandin, who appears to be able to see the world like a livestock animal, seems ill-placed to be designing abattoirs, the places where livestock go to die. But it's exactly her place on the autism spectrum that gives her insight into how animals feel, and to see things as a non-human animal might see them. What she senses when she walks through abattoirs, past gates, up ramps, is what the animals seem to sense. And instead of projecting her emotions onto the animals who will die, the way people *without* autism might, she improves the practices of the abattoirs so the animals are less stressed. Interestingly, the person who probably most understands how farm animals perceive the world isn't the one telling us not to eat animals. Rather, she insists that if we're going to eat them, then we must treat them better.

———·———

If we're going to look after animals better, look after the world better, it could be that we have to forge some uneasy alliances. Animal welfare groups dominated by vegans may well be strong supporters of changes that ethical eaters would like put in place, even if those eaters still want to eat meat. Farm lobby groups, who tend to put the interests of their members before animal ethics, can be invaluable if they feel the pressure of the public to change. Supermarkets are often the largest customers of intensive animal producers, and if the large supermarket chains demand change, it can happen quicker.

The challenge facing those interested in animal welfare is to break the apathy around meat. To motivate people to care enough to change their buying habits. And while graphic images and simplistic messaging may attract intense media coverage, does the usual animal rights campaign of horror and threats actually work? Despite the fear campaigns, the footage of cramped chicken sheds, of pigs biting the bars that restrain them in concrete sheds, of dead or dying animals in intensive farms, people still eat meat. Australians still eat *lots* of meat, more than ever before. So the fear campaign has failed. It has garnered the attention of those at each end of the spectrum, and rallied the troops who subscribe to such things as a matter of course, but it has failed to alter the behaviour of the general public. What it *has* done is vilify those who think differently from each other, causing division, not consensus. I spent two years putting together a documentary for caring meat eaters that those in the industry just ignored. Temple Grandin, on the other hand, has had massive influence because she is a forceful, insightful part of the system, not simply shouting from the sidelines trying to bring it down.

My view is that it's very hard to get people to change behaviour – be they meat eaters, vegans, farmers or just everyday consumers – if we don't work with those who *can* make change, and at the same time try not to alienate those who think differently. Sadly, the discourse around eating meat, and using animals for human purposes, is surrounded by threats. Threats by meat eaters against those who choose a different path. Threats by animal welfare subscribers against those who eat meat. Threats by supporters of intensive farming groups to prosecute those who report animal abuse.

But I have no doubt that change will come. It has to come, to us all.

To those who eat extraordinary amounts of meat, a reckoning is due when the true cost, in health, animal welfare and the environment are considered.

Change will come to the most intensive industries when they realise that what they do, in our name, doesn't have social licence.

Change should come in butcher's shops and restaurant kitchens, as consumers demand better-quality meat, well handled and served without guilt.

And change will come in our households as we learn to cook more meals with little or no meat, and still retain the joy of the table and the fabric of our society.

There is no unbridgeable divide between people, only different starting points. That a vegan was the person who taught me to kill is instructive in that we both arrived at the killing cone through different paths. There will always be differing opinions, different moral codes, different ethical standpoints. We must accept that virtually nobody else will share our exact unique shade of moral ambiguity, and learn to live with that.

It's important that society has the last say on rights and wrongs, and that within this we each make our own conscious decisions based on what we value the most.

And if that includes eating meat, there's a whole bunch of valid reasons to do just that.

CHAPTER 22

Forget the Other Animal for a Second – Look at the Producer

I once wrote a love letter, in the form of a cookbook, to my unborn son. Started when I was unlikely to have any kids, written mostly while single, and finished just after Hedley was born, the idea was to condense everything I knew about food into one volume. A how, why, what manual of all I knew about food.

I had become obsessed with quality. What defines it, who decides it, how it comes to be in the ingredients we buy, the food we consume. And while *The Real Food Companion* spends 60,000 or so words on various topics from grains to lobsters to peaches, I can sum it up in a sentence. The ingredients we buy find most of their differences in the people who grow and produce them. Look at the soul of the producer and you'll get an idea about the inherent quality of the food they grow. When you want to work out why some things are better than others, it's the farmer who is more important than the animal they rear.

So what defines a good farmer? Criticising a farmer is like criticising a parent. There's no one-size-fits-all. But there are clues.

A good farmer views their role in the longer term. A day is only a part of a week, a week only part of a year, a lifetime just one lifetime of many that can be spent tending that one bit of soil. A good farmer knows they are just another in a long line of people, each standing on the shoulders of those who came before, each casting a shadow for those who come after. What we do today, as stewards of the land, has effects that may outlast us, both good and bad.

A good farmer knows what they don't know. In other words, they are humble enough to realise that the systems we deal with are complex. From the billions of microbes beneath our feet in healthy soil, to the plants that grow in that soil, to the ecology of everything around us that relies on nature, all of these things are complicated. Add in weather systems, microclimates, vagaries in soils, seasons, and the personalities of the animals in our care, and it's unbelievably complicated. If you think you know it all, you're wrong.

A good farmer knows a cow isn't a human, and that one cow isn't the same as another. What we have to do is put conditions in place where the animals get to do what they have evolved to do, eat what they were designed to eat, and behave as they were designed to behave, and try not to always see them as other versions of us.

A good farmer is focused on what the end use of the produce is. If it's food, then make it nutrient dense. Make it taste. Use the seasons to ensure what you grow is at its best, not just that it looks good under supermarket lights after spending ten months in cold storage. A good farmer works on the basis that the end use of all things they grow for food is for someone to eat it. The end use is for people to fuel themselves, to feel and spread pleasure, and to gain all the nutrients they need. The end use is cultural, gastronomic and social.

Why do we need to focus on the farmer? Because it's our farmers that can make a difference. They can change an animal's life today. They can start to look at the environment in which they farm in the long term. They can put in place all the components that will make a difference to the land, to animal welfare outcomes, and to the quality of the food that is the end product of all their labours.

Nobody is infallible. We all make mistakes. Understanding animals isn't easy. As farmers, as a society, we try things, they work or don't work, or they could work better, and we learn and improve. Animal welfare, like all things, isn't a static concept, and it can take some trial and error to make things better. And our behaviour also needs to adapt as we understand more, as technology evolves.

As things change, society also needs to understand what is done on our behalf, so we can give social licence to farmers. In the past, what we've done is let farmers do things without our consent, and then been shocked at the results. If you're in the intensive animal industry, and hanging around folk from that same industry, you may not realise that the community demands better, demands more. You look at a farrowing crate where the sow can't move except to stand and lie down, and you see a 'piglet protection pen'. Consumers see a pig jail. You see a computer-controlled, climatically perfect chicken shed with 50,000 birds. The general public see 17 birds per square metre and think it looks a bit, ahem, crowded.

When I think back to the feedlot owner from earlier in this book, and his cattle and his attitude to shade, I suspect that he and I are never going to see eye to eye on some aspects of animal husbandry. The kind of farmer who believes animals don't need shade in a Queensland summer doesn't think like me. The farmer who can run a chicken shed with tens of thousands of birds while arguing it isn't intensive farming is someone I will probably never really relate to.

When I was on a million-acre cattle station, looking down at the ground from a helicopter, I began to get a literal bird's eye view of farming and its impacts. I could see a landscape in flux. Down below the chopper I saw cattle hanging near water holes, hanging near trees – and out in front of me the majesty of the open sky and the big red of Australia stretching out to meet each other on the horizon. From above, you really can see areas of overgrazing. Of open dirt, of damaged soil. This farm, where the water runs eventually into the Great Barrier Reef, had erosion gullies lacing the country. On other properties nearby I found erosion gullies over three metres deep, ravines that towered over my head. For years scientists have warned about the damage that is caused to the Great Barrier Reef by runoff. By fertiliser use, too. And here, where the owners decried the possibility that the law might soon change so they couldn't clear more land of the trees that seemingly held at least some of the soil together, I started to wonder about the damage farmers cause. I started to think about the phrase 'mining the soil', because to me it looked like short-term gain was going to cause really long-term damage.

And I started to think not about farming, but farmers. Of the kind of people who run these farms. The sort of person who reckons there's no money in beef farming unless you run so many cattle that you end up with erosion gullies 2–3 metres deep – well, that ain't the mob I'm used to hanging with. Does it meet community expectations that they should be allowed to clear even more trees to create more grazing, when their farm's within the catchment of the Great Barrier Reef? The farmers I was with stated this was their clear intention, determined to fight land-clearing legislation – before choosing which of their three planes to fly back to Brisbane in. (It was the $2.5 million one, just for the record.)

These are the same farmers who referred to Aboriginal people as 'lazy critters'. Just to put this highly racist slur in context, the owners had just told me how hot it was getting in recent years, out in central Queensland. Almost too hot to live there permanently anymore. They'd talked about increasingly unreliable rainfall. They'd argued that they and their neighbours should be able to cut down even more trees, when the science shows trees help make rain. And then, having expressed their opinions, they buggered off back to Brisbane where they lived most of the time, and criticised the very people who had fed themselves for a good 60,000 years or so off this same land. Aboriginal people who had gathered food for what is an eternity in human history, a period spanning six times as long as modern agriculture, were the focus of blame. Calling Aboriginal people, who could feed themselves sustainably, with virtually no diminution of topsoil, for what is arguably forever in human terms, 'lazy critters'? These farmers, who practise an agricultural system yet to be proven to last beyond a couple of hundred years in these climes, are they entitled to criticise like that? The people who practise a farming system that has turned good land into deserts in some parts of the globe? A system that has ostensibly washed half the farm out to the ocean since its inception?

Are these the farmers you want doing things on your behalf?

I've delved a little into livestock wants and needs in this book, but it's the human animal that drives things most. We're the ones who came up with feedlots and intensive chicken sheds. We're the ones who have spent the last 10,000 years of agriculture, and a few thousand years before that, selectively breeding animals and manipulating landscapes to our needs and wants. And we're the ones who can bring about change. Not just the farmers, not just the producers, but all of us who put fork to mouth.

We're all gifted with the possibility of contributing to social change every time we get out our purse or wallet. It's humans who have created some of the problems we're now encountering, some of the moral dilemmas that now confront us. And it's the human animal that must come up with the solutions to get us out again.

CHAPTER 23

The Need for the Ethical Omnivore

You can almost hear the silence. The noise that we meat eaters don't make concerning animal welfare. For a long time, too long, we've relinquished our responsibilities to those in the game: to those who oppose the use of all non-human animals, and to those who will do whatever it takes to farm animals.

To some, especially those in the more militant animal welfare groups, there's no such thing as an ethical omnivore. But I would argue differently – that it is more important, nay, essential, that meat eaters step up and take charge of animal welfare, because what happens to farm animals happens because we are the ones who actually pay the farmer to farm.

Animal rights, and the choice of some people to eat meat, even as a conscious act, are divisive topics – perhaps understandably given the history and misunderstandings I've dealt with in this book. For too long, though, it's been those who want us to abstain from eating all meat, those who aren't actually having animals reared and killed on their behalf, that have taken the lion's share of load when it comes to upholding community standards.

Despite our sometimes differing views, I'm more than happy to admit that vegans have done wonders for promoting animal rights and the welfare of sentient beings. Peter Singer, one of the English-speaking world's most prominent animal ethicists, founder of Animals Australia, and author of *Animal Liberation* in the 1970s, is vegan. Ingrid Newkirk, the woman who founded PETA, is a vegan. Lyn White, Director of Strategy at Animals Australia, the woman who exposed shocking animal cruelty in Australia's live export trade, is also vegan.

And hence, I see a problem. Not with what these high-profile vegans and their organisations do, but what those of us who have chosen to eat meat *don't* do. And what we don't do, often enough, is stand up for the rights of the animals we use for food, drink and clothing.

So I don't have a beef with vegans talking about animal rights. But I do have issues with meat eaters abrogating their responsibility on the issue of animal welfare. Surely, if we choose to eat meat, to have others breed, rear and kill animals on our behalf, we should have an opinion on how those animals are treated? Isn't it incumbent on the omnivores of the world to have *more* to say about animal welfare than those who choose not to eat animals? After all, it's for our benefit that these beings are brought into this world, husbanded during life and then dispatched at our command.

As I mentioned earlier, this issue hit home for me in a big way when I first stepped from the city to the bush. From the suburbs to the farm. Every decision on Fat Pig Farm has to incorporate not only our family's wants and needs, but also the wants and needs of our animals. Sometimes, I have to admit, we get it wrong. Sometimes, as custodians and guardians of our livestock, we fail. There has been suffering on my farms. We have, as I've said before, unwittingly extended the pain in an animal's life.

The important point isn't that we eat animals, but that there is no *us* without impact on other animals – regardless of whether you eat meat or not. The question is, really, what are we doing to make sure our actions, and the actions of those who farm on our behalf, stand up to scrutiny? We need to think about what our impacts are, whether they meet community expectations, who is responsible for policing them, and at what point on the continuum do you as a person want to place yourself in terms of personal accountability for what happens in your name?

I don't have all the answers, but I do hope I've posed some of the important questions. The reality is that I have made a very conscious decision to eat meat. I know that no matter what I eat, something is affected. I kill more living things (usually slugs and aphids) when I grow and eat cabbage than in a year's worth of beef consumption (though my cows probably ingest the occasional insect or two, too). I am unabashedly guilty of what Peter Singer calls 'speciesism' – a kind of animal discrimination where I favour one or more species over others. But I also think it is incumbent upon me, as someone who eats meat, to take an interest in how, where and why we raise animals. In other words, of all the people who should have *more* motivation and *more* reason to be involved in the discussion, meat eaters are top of the list.

And yet, most of us don't think about the animals we eat. It's up to us, as a society, to decide what is, and isn't, acceptable in terms of how we treat our animals. At present, you can legally house your pigs in a way that you cannot house your dogs. Pretending not to know what happens to animals raised for human uses is not an option, unless you choose not to buy from that system.

Informed, reasoned, vocal opinion from the omnivore is vital, or farmers may do things we don't agree with behind the locked gates or barn doors.

That said, as we've seen, farmers only do things on our behalf. They aren't the enemy, they are merely doing our deeds for us. If we want cheap at any cost, we can have cheap. Just don't open the door on the shed where the chickens are housed unless you want a nasty shock.

If we want farmers to rear animals that have been able to express their instincts, then we as a society must decide that's important – and pay more for these animals if that is what's necessary to achieve that aim. As the saying goes, if you want something cheaper than the real cost of production, then something must suffer: the farmer, the animal, or the environment.

Omnivores need to be part of the conversation. It's no wonder we find it hard to be heard, however. Those with an interest in stopping the world from eating meat, or rejecting calls for open and accountable animal husbandry, have been louder for longer. It's always the radicals who will be motivated more to speak out. There will always be extremists who will go to inordinate lengths for their cause – and perhaps we need them to highlight problems on both sides of the farm gate. But the difficulty here is that extremists of any kind can easily alienate the mainstream. And the mainstream in Australia eats meat.

Animal welfare isn't a fringe issue. It isn't a vegan issue. It isn't even an omnivore's issue. It's a discussion that the whole community should be involved in, because how we rear, house and dispatch other living creatures is a barometer of how civilised a society we really want to be.

CHAPTER 24

What Is Good Meat?

For most of this book I've talked about animals. About farming. About our impacts on other life around us, and on ourselves. But like most individuals, when I think of meat, I think of all the joy that it brings. The flavour, the texture, the cultural relevance and context. It's about cooking and eating, and the pleasure of the table.

One thing I hope you glean from this book is that I think we eat too much meat. More to the point, if you have heard the stories from intensive farms, and seen the impact of our insatiable appetite for cheap meat, and if you have chosen to eat meat, knowing it may have a higher environmental cost than eating only vegetables, and that a beast has died in order for you to be able to eat meat, then surely you want better meat, something of quality, with flavour.

I trained as a chef in the early 1980s, I've been writing about meat for over two decades, and one of the things I've learned is that English is not the most expressive language when it comes to carnivory. Generally we use only two general descriptors and their synonyms to convey meat quality: 'moist' and 'tender'. It's like describing wine as 'cold and wet'. Neither of these words actually talks about flavour.

So what do we describe meat as tasting like? We say rabbit tastes like, well, chicken. And chicken is supposed to be 'bland and tender' according to the industry that rears them. Hardly a recommendation of good meat. Steak can taste, um, beefy. Something strong flavoured is invariably 'gamey'. If you look at a single type of oil, olive oil, the industry has managed to construct a flavour chart that includes lots of descriptors, such as bitterness, grassiness, green apple, woodiness, mustiness and more. Why is meat so hard to describe? Perhaps it's the simple robustness, the chew factor, that means we concentrate less on the flavour of meat than we do of something liquid like oil or wine or cider.

Even professional tasters lack, in English, the words to describe flavours in meat, with some just using the same words as anybody else might, such as 'meaty', or 'lamby'. But there is evidence that, when given a direct comparison, consumers *can* taste better, more flavoursome meat, and it does add to their enjoyment of it.

It turns out there are over a hundred well-researched factors for the taste and aroma in meat, and about 1000 documented volatile compounds that affect meat flavour – and probably thousands of other chemical compounds and their interactions and perceptions that affect taste. The compounds found in meat that are well known to affect flavour include hydrocarbons, aldehydes, ketones, alcohols, furans, pyridines, pyrazines, thriphenes, pyrrols, oxazols, thiazols and sulphurous compounds. Raw, we taste little of these in meat; it's only when heated and served with some warmth still in it that meat reveals its true flavours.

What gives meat its flavour?

Meat is muscle. The more a muscle moves, the more it can grow – and along with more protein, it also lays down lots of those other flavour compounds. There are also two extra major proteins heavily involved in muscle movement, myosin and actin. The more a muscle works, the more myosin and actin are present in and around the cells. Like all proteins, when heated, these break down (denature into other elements or their component amino acids), first myosin at lower temperatures, and at hotter temperatures actin. The breaking down of myosin below about 60°C makes meat juicy – so a rare but cooked steak can taste juicier than a raw steak, strangely. But the actin, when heated, starts to contract at higher temperatures, which forces the moisture from the muscle cells, leading to drier meat. Hence a well-done steak is tougher and drier. In blind tastings, most people prefer meat cooked to a point that the actin doesn't contract. But not all cuts can be cooked medium or rare and still be palatable.

Cooking meat is transformative, but that transformation, and our perception of flavour, does vary widely depending on the cooking method. Meat that is simmered, or poached, undergoes relatively minor reactions, so the flavour is more true to the original meat – albeit with some extra flavour from melted fat and collagen, along with the denatured proteins, and peptides, adding their own zing.

High heat has a much more dramatic effect. When fried or seared or barbecued, proteins and amino acids react with the meat's natural sugars, to produce what's called the Maillard reaction. This reaction, identified over 100 years ago, produces amazing new flavours that aren't like the original component protein or sugars. Rather than the texture of meat, it's the Maillard reaction that really makes the flavour count.

The Maillard reaction happens all the time in cooking. Think of toast compared to bread. Or the muffin top compared to the – go on, admit it – far less interesting muffin beneath. And think about the amazing smell from the searing of meat in a hot pan. It's the Maillard reaction that causes this change, transforming those carbohydrates (sugars) and proteins into something more delicious, more mouth-wateringly wonderful than the original.

Maillard reactions of browning meat are the most striking example, but other reactions can also occur at lower temperatures, such as boiling in a more modest fashion. There's a thing called Strecker degradation at lower temperatures – even ambient temperatures (but far, far slower) – that helps add flavour to meat, but any cooking or long curing is really the key to unlocking meat's flavour potential.

The more proteins there are in meat, including myosin and actin, the more amines are produced, and it is amines that create the *umami* effect – the strangely satisfying taste we get from foods. Umami is the rounded, savoury, full-mouth effect that was first identified in Asia (most notably Japan, where the term originated), and has been used only relatively recently in Western circles when we talk about taste.

To understand what the umami effect is, think not only of barbecued meats, but aged parmesan, mushrooms, tomato sauce. It's the natural flavour that manufactured MSG (monosodium glutamate) is trying to replicate, as MSG and similar compounds are naturally found in relatively high concentrations in tomatoes, seaweed, and fermented goods such as soy sauce, miso, shrimp paste and the like.

So the more a muscle moves, the more flavour it has; more protein, more amino acids, more peptides and other compounds already in the meat, and some transformed by heating, all lead to more flavour. (Remember that chicken meat in the 1940s had about a third more protein than it does today.) Most cooks understand this, knowing a hard-working shoulder muscle is potentially tougher, but also more flavoursome, than the fillet. But the difference in flavour obviously applies to the farming system, too, where an animal that moves more in general has more inherent flavour.

Then, there's fat. Fat is a major carrier of flavour, and some scientists think it's the reason we prefer grass-fed to grain-fed beef and lamb. Grass-fed animals lay down more omega-3 fatty acids – the fat more favoured from a health perspective, and the one nutritionists bang on about in fish. And omega-3 contributes more to the taste of the meat. Omega-3 fats are denser than the grain-fed fats, which are omega-6 heavy, and so grass-fed animals can have more inherent flavour.

Fat also absorbs beta-carotene and other flavour compounds from the food the animal has eaten. Feed a pig salmon and it will taste fishy, and because that flavour is stored in the fat, not the muscle, it's harder to shift it by changing the diet in the week or two before death. There's a direct relationship between the quality and flavour of the fat in a pig's diet, and the flavour (and also the nutritional quality) of the pig itself.

Then, along with protein and fat, there's the connective tissue and its flavour. Dishes where the meat is simmered and not browned (think a good Irish stew) probably rely on the flavour of fat and the connective tissue component, collagen, breaking down, as much as they rely on proteins.

———·———

One problem with meat comes with having one single criterion of quality: tenderness. At some point, our obsession with tenderness has usurped our desire for meat with taste. And the factors that create tenderness are in direct opposition to those for flavour. A young muscle, and a muscle that doesn't do much, is tender, but lacking in flavour. A well-worked older muscle has more flavour, but can be less tender.

That's why intensively farmed animals are encouraged not to move. Not only does lack of movement increase the speed at which they put on weight, an idle animal, in a shed or a feedlot, also has more tender meat. Less flavoursome meat, but more tender. So our predilection for meat we hardly have to chew, or hardly have to cook for very long, means we favour flavourless meat from intensive animal industries.

The problem is, when meat has more flavour, you eat less of it. Tender meat from idle animals is dilute meat; meat that leaves us hankering for more. I always find that meat from older, harder-working animals is far more satisfying, so we don't need as much of it. But the industry just wants us to eat more meat, with all the attendant increased environmental and other impacts of rearing more animals. There's a false economy in buying cheap meat, given that you need to eat more of it to feel satisfied, but it's a game that has worked well in favour of intensive animal industries.

Tender cuts of meat, such as the appropriately named tenderloin (fillet) and loin, are called in the trade 'primary cuts'. Flavoursome cuts, such as neck of lamb, or shin of beef, are called 'secondary cuts'. As a farmer, this term seems offensive to me, because a sheep is a sheep, and each sheep has a neck. A cow is a cow, and each one has four shins. There's no 'secondary' when the animal has all the parts, and each has to be valued for what it is. And a good butcher, and a good cook, should be able to treat each cut with the respect it deserves, extracting the most flavour. There's no cow with ten fillets; in fact, the fillet is about 1–2 per cent of the entire animal. There's no pig with six shoulders (though, to be honest, we do like using pork shoulder for so many dishes, it would be nice if they did).

They are entire animals, which we have chosen to rear for our purposes, so getting the most flavour, the most use, the most joy from each being is a matter of honour for us. And I think that respect should be incumbent on all meat eaters, too.

I'd like to think we could use better words to describe our meat. Does it have a herbaceous note? A mushroom tang? An attractive minerality? Deep, earthy notes, or a luscious sweetness? Can you taste the grass on the beast, as you can with milk when it comes straight from the cow (not when the milk comes from animals fed 6–8 kilograms of grain a day, milk which is then standardised and pasteurised)? Is there a vegemitey character, a sense of umami, or a clean finish? Is there a hint of funk, a note we relish in a truffle and could just as well prize in aged beef, or a primordial tang that lies just on the edge of perception? In flavour is found pleasure, and if we're going to raise animals for meat, to kill sentient beings specifically for the dining table, surely pleasure should be at least part of the defining criteria for what makes meat good.

What is good beef?

Good beef, in my view, is grass fed. Studies have shown that generally consumers think grass-fed beef tastes better. It has more complexity, more character, more going on. In short, it has more flavour. It also suits my animal welfare and ecological standards.

But grass-fed beef can require a bit more chewing, because the animal has wandered around more. Hence, 'dry aging' was invented. This process, which now seems commercially antiquated, involves hanging the quartered beast on a hook in a coolroom for several weeks, to allow the natural enzymes in the meat to break down and tenderise some of the connective tissue, and intensify the flavour. Nowadays, with the advent of soft, grain-fed beef from animals that have been fattened in feedlots, and the ubiquity of butchers who buy in pre-cut, pre-boxed meat, rather than whole carcasses, beef is usually 'wet aged'. What happens is that, because coolroom space is at a premium, most carcasses are only hung for a day or so until just chilled, then the carcasses are boned out. The cuts (fillets, sirloins, blade, silverside and so on) are then put in plastic bags and vacuum sealed, then left to 'wet age' in the bag for anything up to six weeks or even more. Packed in boxes, with the bones already discarded, saves on fridge space, as the boxes can be stacked and shipped straight from the abattoir to butchers. Wet aging is suited to younger animals that don't have a decent fat covering, and haven't spent as much time walking and grazing.

The wet-aging system of young animals works really, really well for the people selling meat, because while both dry and wet aging can make meat more tender, there is no evaporation with wet aging, as the meat is hermetically sealed. Dry aging, on the other hand, can have drastic evaporation – up to 20 per cent weight loss – meaning you get less meat; potentially, there is also more to trim off to get to the nice pieces inside. While this evaporation looks like a down side, for the tastebuds it's a bonus. There's less water in dry-aged meat, so the flavour is more pronounced. (More water in wet-aged meat equals a more diluted taste.)

But wait, there's more. Meat dry-aged for 21 days has also gone through a chemical transformation, turning amino acids – the building blocks of protein – into amines through the action of enzymes. Amines are like nature's flavour enhancer, and add a more satisfying, mouth-filling flavour to the finished meat: umami. Until relatively recently, dry-aged meat, hung for at least a couple of weeks, was fairly commonplace. It's the meat your grandparents possibly ate. And properly done, it is complex, and has the perfect amount of chew, with a luscious, lingering flavour that is quite primal in its attraction. In some circles high-end butchers are now aging meat for 90 days. This may give them bragging rights, but it's excessive; the flavour transformation is mostly complete after

28 days, even though, just like aged cheese, some may like the funky nature of older meat. Just like cheese, however, the outside can become dry, and the flavour tainted, so it has to be discarded.

Wet-aged meat, on the other hand, tends to pick up some slightly 'off' notes from the juices that it sits with in the bag. Those juices fester a bit over time, producing a smell anybody who has opened many of the bags will recognise immediately (and not with relish). When I was a cook, we were told that this aroma just dissipated. It doesn't. At structured tastings, I later learned – too late for my many customers in the 1990s perhaps – that these juices have tainted the meat. But people seem used to it, and after all, they just want tenderness, we are told.

Wet-aging meat has another more insidious result, because the work of cutting up the carcass is done at the abattoir, not the local butcher shop. What this means is that the craft of the butcher – the knowledge of varied cuts, the ability to age different parts differently on the bone in the butcher's coolroom – is becoming a lost art. Most butchers these days are more deft with a bucket of bought-in marinade and some boneless wet-aged meat than they are with a bone saw and the ability to value the carcass as a whole.

What we've gained is soggy, poorly flavoured, tender meat. What we're missing out on is complexity of flavour, and a culture of knowing and valuing every bit of the beast. I encourage you to find and befriend a butcher who still deals with cutting up the beef from scratch. Admire their craft, and they will embellish the meaty meals you can put on your dining table.

What is good pork?

My family and I are a bit biased about the pork we rate highly, because we have free ranged an old breed for a long time now. But we really do think the meat from that breed tastes better.

Older breeds, black breeds, have been bred for thousands of years to taste good. Newer breeds – which are required to be white and fast growing, and can be intensively farmed – are bred for large litters and feed conversion and the ability to be locked in sheds. But they aren't bred for flavour. And what do we on our farm do with pork? Well, we eat it, so we think eating quality is paramount. We are lucky to be in the position of growing food not to sell, but to eat.

Breed and feed are the two biggest drivers of flavour, and we like to think that the animal welfare system, and the chemicals used to get pigs lean and to grow fast, should be part of the bigger picture too. In terms of breed, ours are Wessex Saddleback. We've enjoyed Berkshires, Large Blacks, Tamworths and Durocs, and crossbreeds may well have good flavour profiles, too. There's also a Large White, but I've had limited experience of that old breed. The Gloucestershire Old Spots, extinct in Australia, is a delicious pig if you are ever in the United Kingdom.

So if breed is important, possibly paramount, what about exercise and the farming system? We know from experience that pigs that have been able to dig have more flavour – as do older pigs, slower-growing pigs. After about six months of age, pigs start to put down marbling in the meat, laying down fats in the muscle that adds to the flavour of the pork. Yet most shed-reared, fast-growing white pigs are only about 3–4 months old when slaughtered for fresh meat.

Pigs that enjoy a varied diet have a more complex flavour – as do fish that have a varied diet, or cattle, or chickens. You aren't just what you eat, you are what you eat eats. When fed grain, animals, be they chooks, pigs, cows, or increasingly sheep, put down a lot of omega-6 fat, which isn't as flavoursome, or as good for you as the omega-3 fatty acids laid down in grass-fed meat. And despite being public enemy number one since the 1970s, when some bad science made a bit of lard the devil's work, fat is absolutely essential for pork with flavour.

So, I think better-tasting pork is at least six months old, and it's from older breeds (our preference is black pigs). Better-tasting pork has ranged and exercised, had a pretty damn good diet, and has some fat.

Once you find good pork, you'll be hooked. And, of course, pork can also be turned into a lot of amazing things. This transformation can take many forms, from some overnight curing to cold-fermented salamis and aged prosciutto. And the quickest transformations, the best yielding, the ones that make the most money for producers, are not the best to eat.

Finding good cured pork is probably even harder than finding good fresh pork. Let's take bacon and ham, for instance. British-style bacon and ham are wonderful products born of the necessity to preserve meat before the advent of refrigeration. In traditional British bacon, salt curing and smoking both act as natural preservatives, though now these are used as much for flavour as anything else.

Bacon, made from pork middles – the bit that covers the ribs, including the loin (in Australia) – can be either dry cured in a salt mix, or in brine (a salty water mix). Ham is usually brined, but legs are dry cured to make goods such as jamón and prosciutto.

Brining acts much faster than dry curing, which is why many pork products such as bacon and ham are often 'pumped' – that is, they have the brine *injected* into them for faster and more even curing; in the process, the meat absorbs water, which also makes it weigh more. For ham, it's important that the cure reaches the inside of such a big cut, particularly if it's on the bone, so pumping is usual, even at the artisan level.

More and more often, ham is also tumbled these days. Whole legs, or cuts, are tossed in a thing that looks like a cement mixer (the truck type, in big operations), and spun and tumbled in the briny mixture, the meat massaged so it breaks down the texture and absorbs more liquid. You can increase the weight of your ham substantially by tumbling – by about 50 per cent, or more if you use the right emulsifiers in the cure. It's a clever way to sell water for the price of pork.

Food can be 'smoked' in the old-fashioned way, over smouldering woodchips or sawdust, or it can also be flavoured with a smoke-scented syrup that you buy from a wholesaler. Liquid smoke, made by condensing wood smoke, and then concentrating it, is the article of choice in the industry. I've seen 'double smoked' products that don't look like they've been in a smoker, and plenty that have smoke flavouring and then get a cursory pass through a smoker –which is less about flavour than it is about perceptions. Proper smoked ham and bacon is put into a sealed unit that produces smoke, ideally from shavings or timber, but more and more from pre-packaged sawdust pellets and the like.

This is the ingredient list from a well-known brand of double-smoked ham, which also boasts it is made in Australia from at least 20 per cent Australian ingredients (mostly likely the water):

> Pork (73%), Water, Salt, Acidity Regulators (326, 262), Dextrose (Maize, Tapioca), Emulsifiers (450, 451, 508), Pork Protein (Herb Extract), Sugar, Modified Tapioca Starch (1412), Thickener (407a), Antioxidant (316), Sodium Nitrate (250), Potato Fibre, Citrus Fibre, Smoke Flavour, Canola Oil.

It's an odd mix of phosphates, carrageenan (thickener 407a) and flavour enhancers, along with pork protein to hold water, and strange things like citrus fibre and potato fibre for … well, just because. I really don't know what they're for, because they have no place in a properly cured and smoked leg ham. This is a 'manufactured meat' product, of the kind you'd find in a very round or square shape, hence the large ingredient list. It's the scraps of other meat, bound together with food glues, and sold as ham. But it still says it's ham on the label. Double-smoked, of course, to sound posh.

We'd rather see manufactured meat products that aren't ham *not* called ham – as opposed to cuts of muscle from the leg which *are* ham. And we'd rather hams and bacons that don't go into an actual smoker be called 'smoke flavoured' rather than 'smoked', which implies these goods have actually been, ahem, *smoked*, not just had some tar derivative waved over or pumped into them. *Smoke-flavoured manufactured imported meat product* may not sound appetising, but it is what these products actually are – even if, sadly, it's not how they are sold.

And lastly, the final straw for ham is that instead of it being cooked in a slow oven, oftentimes it's poached in water, to ensure it has heated through sufficiently. In the old days hams were gently baked, or the smoker was hot enough to cook the ham. But now, with the water bath or steamer ovens set on really high humidity, instead of losing moisture, all that extra liquid in the ham stays sealed in – which would be fine except that a large part of the meat you're eating is already diluted with water.

Good pork can be harder to find than it should be. But a proper butcher should have access to well-reared meat, and even many supermarkets now have a free-range option on the shelves. The best way, however, is if you can buy from a grower at a market, because of all the meats, pork is most often misidentified at the retail end.

What is good chicken?

Now, here, I'm probably going to lose you.

As we saw earlier, just about every chicken you buy is of Ross or Cobb genetics: white, fast-growing birds designed for the system that intensively houses and feeds them. Problem is, they taste of little. But if you're happy with bland, soft meat, well, at least let them live a half-decent life.

RSPCA birds are minimum standard. The sheds have to meet criteria around stocking density, litter maintenance (the covering of the floor and how wet/pooey it gets), lighting, exercise and 'enrichment' (something for the birds to do besides eat and sleep, like having a CD hanging from a string to peck at). These are basic standards, but they do at least give the birds more ability to express instincts, and provide better health outcomes.

Even better is certified free range, with even lower stocking densities, and access to the outdoors once the birds are feathered. The chooks only get feathered, though, from about three weeks on, and some birds are already at the cutting shop after about 30 days, so it's not much of a life. Bigger free-range birds will have wandered and dust-bathed and foraged more, however.

Organic chicken is a step above free range, meaning no antibiotics in the feed, even lower stocking densities, spray-free grain and access to the outdoors. The birds grow slower, move more and hence have more flavour.

But best of all is pastured, where birds are moved to fresh pasture in moveable sheds, able to peck at grass, at bugs, and to dust-bathe. And the best of these pastured birds are those of alternative breeds, such as Sommerlad or Joyce's Gold Heritage meat chickens. These birds, scarce in the industry, can take 12–18 weeks to reach full size, which means they've been free to roam on grass for at least eight weeks. But there is a cost. At every step up in animal welfare, the price of the chicken jumps. It's barely a blip from conventional to RSPCA-approved – but a sturdy, flavoursome, pastured chook of a decent breed can cost over $40.

It sounds like a lot, but it's what our grandparents probably paid as a proportion of their income. What I find is that a big, slow-growing bird can cater for at least two more meals for our family than a bland fast-grown chook. We've raised Ross and Cobb chickens, old-breed chickens such as the Indian Game bird, and the Sommerlad birds on our farm, and know the differences well. You do, really, get what you pay for – but what's also worth remembering is that the *chicken* gets what you pay for, too. Sometimes it's worth considering the animals when you're standing at the checkout.

It's also worth learning how to joint a chicken and cooking it on the bone, as this gives even a Ross or Cobb bird more flavour, and you get the bones to cook for stock – an extra meal or two from the life of a sentient being. And that *has* to be worth something.

CHAPTER 25

The True Cost of Food

It's easy to argue that the intensification of animal farming puts food on the average Aussie battler's table, at a price they can afford. By suggesting we eat less meat, or better-quality meat, it's easy to be accused of favouring the rich: perhaps only *they* can afford the grass-fed, organic, free-range alternative? I know only too well how much more these higher-welfare animals can cost to produce.

So let's take a look at the numbers. The average Australian spends about 14 per cent of their income on food – down from about 19 per cent of income 30 years ago. According to government statistics, total expenditure on meat and seafood was only $650 per person in 2015/16, compared to $734 in 1988/89, allowing for inflation (the data for seafood and meat were compiled into one number, unfortunately). We spend less on meat than we used to, and buy more of it. So now, according to the most recent numbers available, each week households spend an average of $13.70 on vegetables and $9.60 on fresh fruit. Compare that to the $40 or more we spend each week on takeaways, fast food and confectionery. Or the 31 per cent of our food budget we spend eating out, a 50 per cent increase on three decades prior. Or the $13 we spend, on average, per household, per week, on our pets.

In short, we've never spent less, as a proportion of our income, on food in general, and meat in particular. Saying we can't afford proper food, when only some of the poorest in the nation may have trouble affording it, is disingenuous. What's more, while up to 50 per cent of what fruit and vegetable farms produce is discarded at or near the farm, about 40 per cent of what we buy at the shops gets thrown away at home, a totally squandered resource.

The truth is, we *can* spend more on food. We used to spend more on food, and historically have never spent less. We can support better farming systems, more sustainable farming, better animal welfare – but most of us just don't, or choose not to. Or perhaps, because for so long farmers have said 'Don't worry' and 'Don't look', we've not fully realised the choices we could make to help make the world a better place.

Farmers don't grow food you want to eat. They grow food you'll buy. They're running businesses, not charities. They don't do it as a hobby. So if you want a farmer to do something, then you need to make it attractive to them in a financial sense. For too long there's been this condemnation of farmers on the one hand for unsavoury practices, and a complaint about the cost of food on the other. Yet as we've just seen, the cost of food is a furphy. Twenty years ago nobody would've believed we'd all find an extra $100 a month for internet and phone access. Today, we just wear the cost.

And if we want better animal welfare outcomes, if we want more sustainable environmental management, if we want great-flavoured meat from animals that have been able to express their instincts, then we'll have to pay for that, too.

It's important to realise that animals do die in our name, whether or not we eat meat – but probably we have more control now as consumers over the lives of the animals that end up as meat. And it's important to recognise that no system is perfect, but if we eat less meat, as a nation that eats so much meat and leads the world in this respect, then we'll probably be doing less overall harm. It's also good to understand that welfare potential is greater in some farming systems than others, but that there's no perfect farmer, just as there's no perfect anything. Higher welfare potential – for instance in a free-range system versus a cage system – doesn't mean that potential is realised, but at least it *can* be realised. In the cage system, however, it never can be.

———————

To make substantive change, there are real issues to overcome. We know that consumers expect better animal welfare, but are motivated by money more than their scruples when they see the price difference at the checkout. In the United Kingdom, which brought in high welfare standards for pigs, the net result was a loss of local pork producers. So it can go wrong.

The UK experience is sobering. In the 1990s, a group called Compassion in World Farming started to campaign for the removal of sow stalls (called gestation crates in Britain). These crates are where the sow lives her whole life, from insemination to just prior to birth, when she is moved to a farrowing crate – a similar cage, except with more room for her piglets (not her) once they're born. Essentially the sow is kept in these crates her entire adult life; wire cages that allow the animal to stand and lie down, but not turn around or express normal behaviours. The crates have to be at least 1 centimetre bigger than her in length and girth, so the sow doesn't have the bars touching her the whole time. Most people, myself included, find the use of sow stalls confronting, the complete denial of motion and natural animal expression too much to bear. But they're really common globally, though the EU has moved away from them in recent years.

The United Kingdom decided to lead the world in ridding their pork industry of sow stalls, an element that didn't meet consumer expectations, in the late 1990s. All the surveys showed loyal British consumers would support the local farmers in that endeavour. But it didn't pan out that way, as other events got in the way. The British pound happened to rise in value, making imports cheaper; there was also a glut of pork in Europe, the United Kingdom's favoured trading zone. Britain suffered from outbreaks of mad cow and foot and mouth disease. The major supermarkets didn't champion British pork growers – and consumers made the most of cheap imported pork meat. In a decade, the number of pig farms in Britain dropped 40 per cent, and instead of producing 80 per cent of their own pork, they now produce only about 50 per cent of demand.

Cheap meat trumps animal welfare once we're at the checkout, and once we're at the dining table, despite all the research pointing to a huge interest in the ethical origins of our food.

There are lessons to be learned from the UK experience. Some of these have to do with attachment. One of the best ways to change behaviour is to remove the simple financial transaction and replace it with an emotional one. So, yes, you'll still have to pay for your beef, but if you've met the farmer, or a farmer like the one you're buying meat off, you are more likely to support them in their endeavours.

Economists know that self-reported surveys about spending habits are unreliable. They know that our behaviour at the checkout has little to do with our value system (the supermarket test, which I talked about earlier) – *unless* we have some emotional attachment to making a better decision. Australian pig farms, like the British ones, could struggle against cheap, poor-welfare imported pork, *unless* they engage with their customers and open their doors. *Unless* they show us that they're doing things we give wholehearted approval to, and gain an emotional attachment to. If farmers hide their operations, we figure we might as well buy meat from an equally appalling system, and just pay less for it.

There's more than just the financial transaction at the heart of our choices. Australians already spend more on alcohol than we do on meat and seafood. We spend not because we have to, but because we want to. We enjoy the flavour of alcohol, and the effects. But over the past few decades, we've started to drink less, but spend more on the alcohol we *do* drink. So it's not just the alcohol content that matters. It goes deeper than that.

It's telling to see how wineries, which are the top end of the alcohol market, have convinced consumers to spend more on their products partly by opening their farms to visitors. Sure, it's more poetic to sit sipping pinot gris overlooking vines than it is to visit a muddy, slightly whiffy pig farm, but wineries have nailed the emotional connection needed to change behaviour. Once you've been to a winery, and enjoyed their produce, when faced with the varied choices in a bottle shop you don't just choose on price. Often, you choose based on your attachment to the winery you've visited.

Animal farmers could learn from the winery experience, where price is only one of the many considerations in place. Consumers want good stories. They want to feel they do little harm. They want to feel that what they do, to feed themselves, has ripple effects that are also as good as their intentions.

The problem for livestock producers in Australia is that secrecy, obfuscation, and operating without a social licence mean nobody has an attachment to any farmer's food.

Australians, as we've seen, are heavy meat eaters, consuming on average just over 110 kilograms each year. This is more than three times the global average, which is a mere 34 kilograms per annum.

It's not out of the realms of possibility for Australians to eat less meat, which would be good for us, good for the environment, and good for the animals we rear. As we've seen, animals die for *all* human activity, but many are born and die at our command, the ones we rear specifically for meat. When you stand at the till, waiting to pay for your next steak, when you pass through the checkout with your pork chops, you're a long way from the origins of your meat.

Some 5000 years ago, we were all farmers. About 5000 years before that, we were all hunter-gatherers. It's for only a blip in history that we've been removed from animals, both their care, and their death. Throughout evolution, humans have spent at least ten times longer as hunter-gatherers than as farmers, and virtually no time separated from the land, the animals and plants until relatively recently. But when you're lining up at the butcher, you're now a long way from what it looks like at a home kill. When you buy plastic-wrapped supermarket meat, you probably feel a lot further from the responsibility that we all once had for our food. In times gone by, we could see the direct result of our actions when we killed, when we grew crops, when we harvested. We weren't removed from the reality of farming or foraging or hunting. Now, thanks to modern cities and more complex societies, we are not only physically removed, but emotionally and morally removed from the world as well – and the animals in our care are the worse for it.

Cheap meat? The community picks up the cost elsewhere, or the next generation will pick up the cost. Antibiotics of 'last resort' are advertised as growth promotants by at least five companies in India. We've lost a fifth of the Amazon rainforest mostly to grow beef, or the soy that is fed to feedlot beef. In the United States, authorities recently recalled 200 million eggs in a salmonella scare. Imagine the birds, the grain, the male chicks, all wasted because of simple industrial farming gone wrong on one week in one country. Salmonella, virtually unheard of in chickens and eggs a century ago, is now the norm.

Choosing good meat has ramifications not only on us, today, but in the long run, too. Our separation from food's origins are doing us no favours, as a species, and not doing other species any real favours, either.

———·———

How much more will a bit of minimum animal welfare cost us? Well, nothing at all if we just eat less meat, like our doctors recommend, and like the rest of the world already does, with the exception of the United States. Not everybody has to have pastured chicken, but we can all do with raising the bar on the lowest standards we accept. Take eggs as an example. A dozen free-range eggs from Coles cost $4.20 for 700 grams when I checked their online shop. Cage eggs, from battery hens, cost $3. Cage-free, one brand at least, was priced at $3.75. Organic Coles eggs cost $8 (though that carton was only 600 grams).

You can see there *is* a price to pay for higher animal welfare. But a dozen cage-free eggs only cost 75 cents more – a paltry six cents more per egg. For most people, it isn't too big a price to pay: 15 cents more per omelette, 18 cents more for a cake. Community standards don't dictate free-range or organic eggs; those are consumer preferences. But do cage eggs actually meet the expectations of the average person in the street? Does the complete negation of any access to normal behaviour (besides breathing, eating, pooing and laying) justify the meagre savings in such a wealthy nation? I don't think so, but it's not up to me. It should be up to all in society, not me, and not the battery-hen farmers alone. Not the animal rights campaigners either. It should be the egg eaters who are also given voice – the strongest voice – in this debate.

Good, ethical meat and animal products should stack up to community standards. We need a moral baseline for the animals that are in our care, a baseline that extends across the farming divide right into the domestic kitchen. It's important that some of these decisions are made before we head out to the supermarket, before we sit down to dinner.

Because we've changed our association with the animals we use for food over the last few centuries, and the last few decades in particular, we need to change our behaviour. It's time to reframe how we look at farming animals, what we see as their needs and wants, and how to feed ourselves while not buggering up the world that supports us in the process.

I see the approach as multi-tiered.

First, legislators need to realise that community standards are vital for trust in the food system. If cage eggs from battery hens don't meet that standard, then let's make that system illegal. Ban it, and then we'll know that all the laying hens in Australia will meet minimal consumer expectations.

To make changes across all farming systems, we'll also need truly independent science that looks at what are ideal outcomes for animals (apart from, well, you know, their eventual death at our hand).

Legislators should look further afield, too. If it's not okay to rear an animal a certain way in Australia, products from overseas that don't meet our national ethical standards shouldn't be allowed in either. This might mean some serious thinking around free-trade agreements, but it's disingenuous to pretend we care about animal welfare here, then import meat products from overseas that don't match those standards. If we do allow those imports, we haven't improved the life of a single animal – just exported jobs and profits. We've favoured economics over animal welfare in a way that doesn't even help build our own communities.

I also think the big customers should take their role seriously. Most Australians buy meat and animal products from our biggest supermarkets: Coles, Woolworths, and increasingly ALDI and IGA. These supermarkets should lead from the front, ensuring that what they source meets society's minimum ethical standards. Coles did it with sow stalls, but only for meat under their brand. They could insist on it for all the brands sold in their supermarkets. Woolworths, a less notable institution in animal welfare terms, could do their Fresh Food customers a favour by concentrating on only providing meat that has some open, fair and accountable animal welfare guidelines. Some 70 per cent of Australians will probably buy the ingredients for dinner from these two supermarkets most nights of the week. And those customers don't have the time to read all the labels, don't have the space in their lives to do the research on what each standard means, and they do trust the retailers to do the right thing on their behalf. I've seen, first hand, how the actions of supermarkets have changed animal welfare for the better in feedlots, chicken farms and piggeries. We just need to get the supermarkets to enforce criteria across the board, so that every time you buy things at the grocer you'd know they meet a certain standard.

The big customers should also include takeaway firms. The clout a fast-food chain like KFC could wield, if it chose, in driving the big companies to produce legitimate free-range poultry is enormous. One word from them, and our two biggest chicken companies,

Baiada and Ingham's, who produce 70 per cent of chook meat in Australia, would suddenly take notice. McDonald's could do the same, and also laud their credentials with free-range eggs, as they did with the abolition of cage eggs in late 2017. As the owner of Guzman y Gomez, the Mexican takeaway we met earlier, says, fast food doesn't have to be bad food.

There are all sorts of medium to large businesses, from hotel chains to catering companies, from casinos to high-profile restaurants, that can drive change by the purchasing power they wield. All of those companies are influenced not only by those who use their services, but also by those people who work there, and those workers' moral compasses.

While big companies can drive change quicker, it's actually the little guys – you and me – that can effect change in the long run. Every time you ask the hotel buffet if the eggs are free range, if the pork is sow stall–free, you make it known you value higher welfare standards. Every time you invest a few more cents in an RSPCA chicken, or grass-fed corned beef, you wield economic power. While it should be up to regulators to ensure basic community expectations are met, and that minimum standards apply to all our meat, dairy and eggs, it's individual choices that can drive standards up generally. It's our voice, our expectations, our desire to have meat that we believe isn't doing unnecessary harm that has driven Coles and McDonald's to make the changes they have. If they want our support, if they want social licence, they will continue to raise the bar.

Farmers do things on our behalf. Nobody has 50,000 chickens in a shed for personal use. Rivalea doesn't have 45,000 sows, each giving birth to an average of 26 piglets a year, because they like pigs. They're a business. They're in the game to make money, not make pigs happy. If they make the same money, or more money, meeting higher animal welfare standards, they'll be willing to oblige.

We can't expect big companies to operate on a moral framework; the very nature of the stock market and corporate governance actively works against increased animal welfare.

But we can get those companies to change if we, as a society of mostly meat eaters, demand it of them. If they make the same money, raising fewer pigs or chickens, in a higher welfare system, then it's a win all around. Better for the animals, better for the workers on those farms, better for the environment, and better for those of us who want to believe we should be able to eat good-quality meat while decreasing the amount of suffering that happens in our name.

But mostly, it does cost something to raise the bar. Unless we pay the right amount, the animal, the farmer, the environment – or all three – may suffer. If you want to sleep better at night, you might have to pay a little more.

You might have to pay the true cost of rearing animals, or accept that your meat eating does in fact add to the harm done in the world.

CHAPTER 26

Force for Good

I am at the end of my journey. For over five years I've delved into this complex topic in a serious way, and that's at the end of decades of slightly obsessive thought about food.

I've been to farms that offend not just my sense of smell, but also my sense of right and wrong. I've seen the good, the bad and the truly ugly when it comes to meat production around the country and around the world. For months I've been trying to unravel my feelings about the animals in my care at our farm, and those on farms around the nation, and make sense of our strange relationship to them. Tonight, I've just come from a Slow Meat symposium, where complex topics around ethical food production have been dissected with butcher-like precision, and with an earnestness that would make a vegan convention look like a comedy night.

And now, with a deadline looming, I'm in Melbourne. This is where I first became a food communicator full time, living as a café reviewer, restaurant critic, guidebook editor, recipe writer and product sampler for the newspaper, two decades ago. Ensconced in a hotel for two nights so I can escape the distractions of working from home, and focus on finishing this book, my mind is full of thoughts of big farms and big flu and big topics for the ethical eater.

Of course, there are also many distractions in a big town, so tonight, despite (or perhaps because of) feeling the pressure of a deadline looming, I opt to go out for dinner rather than face the computer screen for one more hour. As I leave my hotel in Little Bourke Street up the top end of Chinatown, I ponder turning left to visit one of the famed noodle bars that dot the street. But I fear for the pork of unknown origins that I'd have to eat. And so I choose to turn right, to head uphill to visit The European, a diner famed not only for its timeless feel, but also for its wine, of a style and quality that I rarely get at home. And faced with a menu that doesn't give the origins of the duck, the pork, and worn down by my own pedantry on meat, of course I choose the vegetarian option, of pasta filled with creamy burrata.

Around me are people just flown in from overseas, getting their first, long-awaited fix of their favourite Melbourne restaurant. There are people flirting on a first date, people getting takeaway coffee, a couple breastfeeding their baby. There are work groups and the sound of laughter and conviviality and robust discussion and gentle ribbing filling the air. It is, in so many wonderful ways, exactly what heaving, diverse, vibrant humanity looks and sounds like.

Nobody (except me!) cares that the menu doesn't describe the pork in the charcuterie as free range, despite the high price tag. Nobody asks the questions I would ask about meat as they order. This isn't the time or place to bring up the grim realities of food production and animal husbandry.

Sitting here, where the hubbub of good cheer and the rich grease of social interaction are smoothing relationships and binding lovers, it really makes me think. If you don't know when you're spending $42 on a main course whether it's ethically dubious meat, when is it time? If an established diner, where the service and the wine and the food are exemplary – if *they* don't give you information about where their product is from, should we all really care? I can guarantee you that the buoyant mood in the room wouldn't have been helped with a few barbecue bores like me grilling the waiters on exactly which side of the hill the grass-fed British White cattle had grazed on and for just how long.

Perhaps the place and time for this conversation is, rightly, before we go out to eat. We should, as a society of meat eaters, have this discussion about what is good, bad or passable about meat production long before we hit the supermarket or the café. The role of food is to fuel our bodies, but more so, it is the role of food to build culture. To help ease our way in relationships, to bring joy and hope, and cement families and communities. And that is what matters when we sit down to eat. It's not about feeling guilty, or being angry at the food choices of others. Sharing food is about more than just the necessary act of getting enough carbs or protein, but also allowing us to hear each other's stories, to unite humans through a pleasurable collective experience.

So when should this conversation about food happen, and just how do you choose what is good meat, and how do you get it?

If there's one thing I hope you get from this book it's the freedom to understand personal choice and feel no guilt. Being Jain and saying you do no harm by eating no animal products, and no root vegetables, and not playing cricket on grass, does not mean you do no harm. You simply do different, and possibly less, harm. As we've seen, being vegetarian may mean fewer animals die, but it won't guarantee you don't have blood on your hands. And eating meat *can* be a conscious choice for the ethical eater.

It's no good beating yourself up over the meat you eat. Most of us in Australia could probably eat better-quality meat less often, a win for both the animals and our health. Yes, if you have the chance to buy from a farmer you think does a better job, please, by all means take it. But for so many people, for so much of their lives, there's little or no choice. What is done in our names by farmers who aren't proud of what they do, who are fearful of what you might think of what they do ... well, this happens anyway. Not everybody has the geography, the finances, the knowledge to make better choices every meal of every day. And if you don't have the choice, if you're out to eat with a lover, a friend, a workmate, what matters in that moment may be more important than the origins of the chicken that you have chosen to eat. In that moment, perhaps it's better to choose the least harmful option, which could be the vegetarian option, or it could be the lamb. But nobody should feel ashamed if in that moment they do prioritise other things in life, and order the chicken.

The important moment isn't when you *don't* have the choice. The important time is when you *do* have the choice. When you've got the money for pasture-raised chicken. When you've got access to the farmer who can show you photos of their truly free-range pork. When you can visit the butcher who can tell you the origins and the ethical motivations of the farmers who rear the animals on your behalf, or the fishmonger who caught your sardines, or who can give you the name of the boat and skipper that caught it. That's the time when you have the chance to change the world.

Every time you can make a better decision, that is the power you have to make real change. Each farmer who does things less for the money or more for the animals in their care, they are motivated and energised by you, the people who choose to buy from them.

We already incarcerate animals in a fashion most people in our society wouldn't approve of if they were allowed behind the locked gate. We already use chemicals with dubious long-term outcomes that may cause problems in human health. We already strip animals of the freedom to express their instincts. Going out to a restaurant and being part of that system changes nothing. But it's also not worth anguishing over past choices or poor choices.

Your power comes not at that time, but when you do have the chance to make a better decision. It's in that moment, standing at the chicken section in the supermarket faced with an organic chook that is more expensive than the $5 conventional one next to it. That's when you have the ability to make a better world, the opportunity to alter the momentum.

Sure, I want meat eaters to get involved in this discussion. I think the government should realise that 'stakeholders' in the intensive meat industry inevitably includes the meat eaters in society – and it's *their* interests that regulators and those responsible for leadership should include, before the farming and vegan lobby.

One day, I hope the baseline for animal welfare, for additives to their feed, will be founded on what we as a society want as the community norm, not something foisted on us because of our ignorance or absence from the debate.

I foresee a time when the worst meat will be a minor part of the market that even then matches the least of consumer expectations, rather than the worst farming producing most of the meat. That only happens now simply because the industry is conducted in secret and thinks we don't give a damn. The reality is, we do give a damn. We just have to show them that we do.

I want those who care about our meat to get involved in the conversation – to start, not end, the discussion of what good meat looks like. I've written not for all the times when you can't make a better choice, but for all the times you *can* make a difference – to your table, your body, the animals impacted on your behalf, and to the environment that nurtures us all.

Eating meat isn't for everyone. But some meat, good meat, is probably not a bad thing to have in our diets. It's been that way since the dawn of time. The difference is, for the first time, we now have to make conscious decisions about how our diet impacts the world, and all its living beings, livestock and wildlife alike.

Now is the time for meat eaters to take responsibility for what happens in their name by farmers, and to animals, around the world. I just hope we're big enough, and have the awareness enough, to make that happen.

ACKNOWLEDGEMENTS

A book like this is the sum of many parts, the genesis as complicated as the topics it covers. The inspiration was born out of the questions that came up while filming the documentary *For The Love of Meat* for SBS. So a huge thank you to the team at Essential Media and SBS for throwing such energy into that project. I'm especially grateful to the researchers on the team, and to Josh Martin who shared part of the journey and enough of the angst. Particular thanks must go to director and producer Steve Oliver who worked tirelessly on the show to demystify meat production. Our endless discussions on whether meat eating is ethical, and who should decide how animals are treated, kicked off *On Eating Meat* in the first instance. Unfortunately, Steve, I'm now a bigger barbecue bore than you could ever be.

I am indebted to my steadfast publisher at Murdoch Books, Jane Morrow, for showing such enormous faith in the project when it was little more than a few thoughts on a single page, and giving guidance when most needed in the writing phase. See, I did get the right number of words, and eventually I did get them all in the right order.

To Justin Wolfers, editor Katri Hilden and the rest of the team at Murdoch, thank you for the careful attention to detail, and being such fans of the topic. It helps to know that there's so much support for a book that asks the tough questions of our high reliance on meat.

To our crew at Fat Pig Farm, thank you for taking up the slack when I've been distracted, immersed in scientific reports, crafting arguments, or chasing statistics, when there's cows to milk, gardens to tend, lunches to cook.

I'm grateful to all the researchers, journalists, thinkers and doers who came before, many of whom are listed in the reference pages. But I'm also thankful for all the discussions I've had with the vegans, vegetarians, slaughtermen, butchers, hunters, omnivores and everybody else who have helped form my views. To those farmers who did allow me to visit your farms to see how you rear animals for the table, my heartfelt appreciation, too.

Most gratitude, as in most things, must go to Sadie Chrestman, my lover, my critic, my sounding board, and my muse; without your unwavering support I would never have made it past the first two chapters.

And to Hedley, our son, whose innocent world view as he has grown has been revelatory. I hope the world you inherit understands and respects its animals better than the one I was born into.

RESOURCES

INTRODUCTION **Smells Like Money**

Page 7 Ammonia in pig sheds:

https://pdfs.semanticscholar.
org/6037/645d5ee02366a0460cf6b7a4a4ecda48055e.pdf

https://www.usab-tm.ro/vol10bMV/54_vol10b.pdf – 104 ppm

https://www.safeworkaustralia.gov.au/system/files/documents/1705/workplace-
exposure-standards-airborne-contaminants-v2.pdf – safe level is 25ppm

https://digitalcommons.unl.edu/cgi/viewcontent.
cgi?article=1107&context=coopext_swine – prevent health risk for both
pigs and humans if it's below 10ppm

Page 11 In Meat & Livestock Australia's 2018 state of the industry report,
the 2017 figures for beef and lamb were 26 kg and 9 kg respectively, with
pork about 26 kg, too. The ACMF says chicken meat is 49 kg – which
makes 110 kg in total (with a tiny bit of mutton, goat, duck etc making
up the difference). The global average meat consumption is just 34.1 kg
per annum. In 2015, Australians ate four times more beef and veal than
the global average, and six times more lamb and mutton, according
to the 2017 Meat & Livestock Australia state of the industry report.

http://rmac.com.au/wp-content/uploads/2018/09/SOTI18.pdf

http://australianpork.com.au/wp-content/uploads/2017/10/Annual-
Report-2016-2017_Final-web.pdf

www.chicken.org.au/facts-and-figures/

CHAPTER 2 The View from the Feedlot

Page 23 Feedlot cattle; stats on supermarket share, etc:

www.feedlots.com.au/industry/feedlot-industry/about

Page 24 Acidosis in feedlot cattle:

US numbers, feedlotted for 140 days or so, 37.8% at finishing stage:

https://academic.oup.com/jas/article-abstract/92/7/3053/4702275?redirected
  From=PDF

Page 25 90% of feedlot cattle put on antibiotics at some stage:

www.mla.com.au/research-and-development/search-rd-reports/final-report-
  details/
  Animal-Health-and-Biosecurity/A-survey-of-antibacterial-product-use-in-
  the-Australian-cattle-industry/2660

CHAPTER 3 Killing the Mothers: The Australian Pig Industry's Dark Secret

Page 33 Sow euthanasia for 'minimal disease' pigs – originally at the
  following link:

www.daf.qld.gov.au/business-priorities/animal-industries/pigs/pig-health-
  and-diseases/disease-prevention/minimal-disease-pigs

Page 33 New link referencing hysterectomy:

www.daf.qld.gov.au/business-priorities/agriculture/animals/pigs/health-
  diseases/prevention/reduce-disease-risk

CHAPTER 4 160,000 Deaths: A Medium-Sized Operation

Page 44 *Sydney Morning Herald* article on tender, moist chicken meat
  that doesn't taste of anything:

www.smh.com.au/entertainment/the-best-chook-to-cook-20101129-
  18dml.html

Page 45 Chicken meat research paper in *Nature*:

www.nature.com/articles/s41598-018-22004-6

Page 48 Feeding poultry litter to cattle:

https://extension2.missouri.edu/g2077

CHAPTER 6 Where are the Animal Lovers?

Page 62 Anna Krien, 'Us & Them: On the importance of animals',
  Quarterly Essay, March 2012.

Page 62 Charlotte Wood's *Good Weekend* article, 'This dog is not a human being... (right?)', 24 September, 2011.

www.charlottewood.com.au/uploads/2/8/8/3/2883343/animal_people.pdf

Page 63 RSPCA figures on pet ownership:

https://kb.rspca.org.au/how-many-pets-are-there-in-australia_58.html

CHAPTER 7 Cats. Beyond the Cute Videos

General statistics on numbers of feral cats in Australia:

Australian Government Department of the Environment and Energy, 'Frequently Asked Questions: Tackling Feral Cats and Their Impacts':

www.environment.gov.au/system/files/resources/bb591b82-1699-4660-8e75-6f5612b21d5f/files/factsheet-tackling-feral-cats-and-their-impacts-faqs.pdf

'Fact check: Are feral cats killing over 20 billion native animals a year?' RMIT ABC Fact Check, www.abc.net.au/news/2014-11-13/greg-hunt-feral-cat-native-animals-fact-check/5858282

Page 68 20 out of 30 mammal extinctions are due to cats:

www.smh.com.au/national/war-on-feral-cats-australia-aims-to-cull-2-million-20170214-gucp4o.html

Page 68 According to the CSIRO, predation by cats is reported for 258 Australian reptile species (about one-quarter of described species), including 11 threatened species:

www.publish.csiro.au/WR/WR17160

Page 69 The Australian Wildlife Conservancy estimates 75 million native animals are dying each night:

www.australianwildlife.org/media/27964/AWC-Wildlife-Matters-Summer-2012-2013.pdf

Page 69 Cats kill a million birds in Australia every day:

www.abc.net.au/news/2017-10-04/cats-killing-one-million-birds-in-australia-every-day-estimates/9013960

Page 69 Toxoplasmosis and its effects on marsupials:

https://theconversation.com/toxoplasmosis-how-feral-cats-kill-wildlife-without-lifting-a-paw-32228

Page 70 PETA's view on feral cats:

What can you do? Turn vegan is one of them:

https://www.peta.org.au/issues/feral-cats/

Page 71 Article on culling feral cats

Julie Power, 'War on feral cats: Australia aims to cull 2 million', *The Sydney Morning Herald*, 17 February 2017:

www.smh.com.au/national/war-on-feral-cats-australia-aims-to-cull-2-million-20170214-gucp4o.html

Page 71 An Animals Australia rant on why feral cat killing is a witch hunt: Frankie Seymour, 'The Great Feral Cat Con Job: The ungentle art of scapegoating and scaremongering':

www.animalsaustralia.org/documents/essay-feral-cat-con-job.pdf

CHAPTER 8 A Farmer's Musing on Suffering

Page 83 2010 Newspoll, 'A Pound of Flesh': One interesting finding is that 99% of Australians are against cruelty to animals, yet 98% eat animals and fish that come from cruel, intensive farming practices! This quote is from the Vegetarian Victoria website:

https://vegvic.org.au/statistics/

www.voiceless.org.au/sites/default/files/PoundofFlesh220310.pdf

CHAPTER 9 Floyd and What We Don't Know About Animals. Yet.

Page 94 General feedlot information from the industry website:

http://feedlots.com.au

Page 94 Farrowing pens: State of the pig industry (NSW DPI 2015):

www.dpi.nsw.gov.au/__data/assets/pdf_file/0011/578747/pork-industry-overview-2015.pdf

Page 95 Farrowing crates rebranded as 'piglet protection pens':

https://aussiepigfarmers.com.au/pigs/our-housing/piglet-protection-pens/

Also check out the shots of indoor housing (etc.) on this site used to promote intensive pig farming

CHAPTER 11 Think Being Vegan Hurts No Animal? Think Again.

Page 109 University of NSW Professor Mike Archer's stats and comments on rodents and other killing that happens in grain and pulse agriculture:

http://theconversation.com/ordering-the-vegetarian-meal-theres-more-animal-blood-on-your-hands-4659

Page 109 Western Australian wheat field stats:

www.agric.wa.gov.au/grains-research-development/western-australian-wheat-industry

Page 109 Mouse plagues every four years or so:

www.nationalgeographic.com.au/nature/the-bizarre-mystery-of-the-aussie-mouse-plagues.aspx

Page 110 200,000 NSW native ducks shot over five years to protect rice crops:

www.smh.com.au/national/nsw/200000-native-ducks-shot-as-pests-of-rice-crops-20140802-zzrjj.html

Page 111 Commercially we rely on bees as pollinators for carrots:

http://beeaware.org.au/pollination/pollinator-reliant-crops/vegetables/

Page 111 More reading on Californian almonds and honeybees:

https://abcnews.go.com/US/growing-california-almonds-takes-half-us-honeybees/story?id=52265334

www.popsci.com/almond-milk-could-be-bad-for-bees#page-3

www.scientificamerican.com/article/migratory-beekeeping-mind-boggling-math/

Page 111 Bugs we eat (0.5–1 kg/year!):

https://blogs.scientificamerican.com/but-not-simpler/i-hate-to-break-it-to-you-but-you-already-eat-bugs/

Page 112 Tasmanian roadkill statistics; 32 animals die every hour:

Ann Jones, 'Welcome to Tasmania, the roadkill capital of the world', ABC Radio National, 14 December 2015, www.abc.net.au/radionational/programs/offtrack/tasmania-roadkill-capital-of-the-world/7021816

Calla Wahlquist, 'Tasmania launches roadkill campaign to reduce 500,000 native animal deaths', *The Guardian*, 18 January 2018, www.theguardian.com/australia-news/2018/jan/18/tasmania-launches-roadkill-reduction-effort-to-prevent-500000-wildlife-deaths

Page 113 Maria Island cull:

www.parks.tas.gov.au/file.aspx?id=24476

Page 113 70,000 birds killed to fly planes into New York:

www.theguardian.com/world/2017/jan/14/new-york-birds-killed-airport-miracle-on-hudson-sully

Page 113 10,000 birds killed to protect Amsterdam airport:

www.independent.co.uk/news/world/europe/amsterdam-s-schiphol-airport-allowed-to-resume-goose-cull-in-bid-to-prevent-bird-strikes-8632906.html

Page 113 Perth airport ibis cull:

www.perthnow.com.au/news/wa/council-to-cull-100-australian-white-ibis-birds-due-to-perth-airport-flight-fears-ng-b88768861z

Page 117 *Wired* article that includes the energy use and carbon emissions of fertilisers made with fossil fuel:

www.wired.com/2016/05/chemical-reaction-revolutionized-farming-100-years-ago-now-needs-go/

Page 117 In-depth reading on fossil fuels used in fertiliser and the impacts:

www.ncbi.nlm.nih.gov/pmc/articles/PMC2935130/

Page 117 Interesting facts on fossil fuel use to produce fertiliser:

www.resilience.org/stories/2006-06-11/implications-fossil-fuel-dependence-food-system/

Page 117 One of the least energy-intensive ways to produce food is to have animals on pasture:

https://artofeating.com/the-true-cost-of-food/

Page 118 More information on the horseshoe crab and its ability to detect toxins:

www.iflscience.com/plants-and-animals/how-horseshoe-crab-blood-saves-millions-lives/

https://en.wikipedia.org/wiki/Horseshoe_crab

CHAPTER 12 Beef Is Killing the World. What About Asparagus? Or Golf?

Good graphics on food systems greenhouse gas emissions:

https://ccafs.cgiar.org/bigfacts/#theme=food-emissions.

Agriculture accounts for about 16% of emissions in Australia:

www.agric.wa.gov.au/climate-change/how-australia-accounts-agricultural-greenhouse-gas-emissions

A good reference for UK carbon (one-fifth of greenhouse gas emissions come from agriculture):

https://link.springer.com/article/10.1007%2Fs10584-014-1169-1

The Food Climate Research Network has very accessibly written and comprehensive research on greenhouse gas emissions from ruminants: www.fcrn.org.uk/sites/default/files/project-files/fcrn_gnc_report.pdf

Page 120 United Nations 2006 *Livestock's Long Shadow* report: www.fao.org/docrep/010/a0701e/a0701e00.HTM

Page 121 The 2014 Oxford study, reported in *The Guardian*:

'The study of British people's diets was conducted by University of Oxford scientists and found that meat-rich diets – defined as more than 100 g per day – resulted in 7.2 kg of CO_2 emissions. In contrast, both vegetarian and fish-eating diets caused about 3.8 kg of CO_2 per day, while vegan diets produced only 2.9 kg. The research analysed the food eaten by 30,000 meat eaters, 16,000 vegetarians, 8000 fish eaters and 2000 vegans.'

www.theguardian.com/environment/2014/jul/21/giving-up-beef-reduce-carbon-footprint-more-than-cars

The original report is here, where the authors state 'We did not account for the cooking process (either at the industrial stage or at home) for any of the food codes':

https://link.springer.com/article/10.1007%2Fs10584-014-1169-1

Page 121 Livestock accounts for 9% of anthropogenic greenhouse gas emissions according to the *Livestock's Long Shadow* report (see above), so getting rid of all meat eating would only remove some (not all) of those greenhouse gas emissions (unless you just killed all the animals).

Also see: www.globalagriculture.org/report-topics/climate-and-energy.html

Page 121 David R. Montgomery, *Dirt: The Erosion of Civilizations*, University of California Press, 2007.

Page 125 How the rumen works:

www.publish.csiro.au/ebook/chapter/SA0501041

www.milkproduction.com/Library/Scientific-articles/Animal-health/Rumen-Microbiology/

Page 125 Numbers of microbes in the rumen:

www.purinamills.com/cattle-feed/education/detail/rumen-microbes-in-cows-101

Page 125 Kangaroo methane emissions:

www.csmonitor.com/Science/2015/1105/How-gassy-are-kangaroos-actually

Page 125 Monogastric herbivores and feed inefficiency compared to ruminants:

http://animalbiosciences.uoguelph.ca/~gking/Ag_2350/nutrition.htm

https://besjournals.onlinelibrary.wiley.com/doi/
full/10.1046/j.1365-2664.2002.00693.x

Page 125 Warming potential of methane:

www.scientificamerican.com/article/how-bad-of-a-greenhouse-gas-
is-methane/

www.factcheck.org/2018/09/how-potent-is-methane/

www.carbonbrief.org/guest-post-a-new-way-to-assess-global-warming-
potential-of-short-lived-pollutants

www.epa.gov/ghgemissions/understanding-global-warming-potentials

Page 125 Methane is four times more damaging over 100 years than
equivalent carbon dioxide emissions:

www.onegreenplanet.org/animalsandnature/methane-vs-carbon-dioxide-
a-greenhouse-gas-showdown/

Page 128 Amazon rainforest clearing:

www.theguardian.com/environment/2018/oct/26/soy-destruction-
deforestation-in-argentina-leads-straight-to-our-dinner-plates

www.reuters.com/investigates/special-report/brazil-deforestation/

www.fao.org/docrep/ARTICLE/WFC/XII/0568-B1.HTM

https://globalforestatlas.yale.edu/amazon/land-use/cattle-ranching

www.independent.co.uk/news/world/americas/brazil-amazon-rainforest-
jair-bolsonaro-deforestation-global-warming-a8649976.html

Page 128 Some Queensland tree-clearing stats, looking at 1 million
hectares over four years:

www.abc.net.au/news/rural/2017-05-22/tree-clearing-debate-re-emerges-
in-queensland/8547116

www.theguardian.com/environment/2017/oct/05/alarming-rise-in-
queensland-tree-clearing-as-400000-hectares-stripped

www.brisbanetimes.com.au/national/queensland/queenslands-controversial-
land-clearing-laws-to-be-overturned-20151128-gladle.html

www.sbs.com.au/news/labor-to-reintroduce-tree-clearing-laws

www.smh.com.au/environment/conservation/revved-up-land-being-
cleared-at-the-rate-of-14-football-fields-a-day-in-northern-nsw-
20180906-p5026l.html

Page 129 Meat & Livestock Australia's Good Meat website:

www.goodmeat.com.au/supply-chain/on-farm/environmental-sustainability/

Page 130 Water footprint references:

https://waterfootprint.org/media/downloads/Hoekstra-2012-Water-Meat-Dairy_2.pdf

https://waterfootprint.org/en/water-footprint/product-water-footprint/water-footprint-crop-and-animal-products/

Greg Peters et al., 'Accounting for water use in Australian red meat production', https://rd.springer.com/article/10.1007/s11367-010-0161-x

Page 133 Simon Fairlie, *Meat: A Benign Extravagance*, Chelsea Green Publishing, 2010.

A well-argued, well-researched view on meat and its impacts, including how the world might look if it did go vegan.

Page 135 Louise Fresco, 'Can Livestock Production Meet the Growing Demand for Meat in Developing Countries?', www.worldfoodprize.org/index.cfm/88533/18099/can_livestock_production_meet_the_growing_demand_for_meat_in_developing_countries

Page 137 Lawns, Australia's most irrigated crop:

www.smh.com.au/national/nsw/the-australian-lawn-short-front-and-sides-20151001-gjz13p.html

Australia's love of lawns harming the environment, researchers warn:

www.abc.net.au/news/2018-12-01/lawn-porn-and-australias-love-of-green-grass-harming-environment/10570548

Page 137 US cropland, corn or lawn?

https://scienceline.org/2011/07/lawns-vs-crops-in-the-continental-u-s/

Page 137 Global land used for grapes to make wine:

www.bkwine.com/features/winemaking-viticulture/worlds-grape-growing-vineyard-acreage/

Page 137 Hectares under grape vines:

www.oiv.int/public/medias/5479/oiv-en-bilan-2017.pdf

CHAPTER 13 Antibiotics: How Farming Is Ruining it for Everyone

US Food and Drug Administration's stats on antibiotic use in food animals:
www.fda.gov/downloads/ForIndustry/UserFees/AnimalDrugUserFeeActADUFA/UCM588085.pdf

A handy visual of how antibiotic resistance can occur:

https://apua.org/science-of-resistance

Other scary antibiotic news: 'Between 2010 and 2030, recent modelling suggests that the global consumption of antimicrobials will increase by 67%, from 63,000 tons to 105,000 tons. For Brazil, Russia, India, China and South Africa, the estimated increase in antimicrobial consumption is 99% – up to seven times the projected population growth in this group of countries.' Peter Collignon and Andreas Voss, 'China, what antibiotics and what volumes are used in food production animals?', *Antimicrobial Resistance and Infection Control*, 2015, volume 4, www.ncbi.nlm.nih.gov/pmc/articles/PMC4415312/

Antibiotics used in Australian agriculture from 1999 – only one-third of the 700 tonnes imported went to humans. Since then, the antibiotic avoparcin has been withdrawn, but was used in pigs, beef cattle and chickens as a growth promotant until about 2006:

www.abc.net.au/science/slab/antibiotics/agriculture.htm

Page 139 Rob Wallace, *Big Farms Make Big Flu*: *Dispatches on Influenza, Agribusiness, and the Nature of Science*, Monthly Review Press, US, 2016.

Page 139 World Health Organization reports that in some countries, 80% of antibiotics are used in livestock:

'Stop using antibiotics in healthy animals to prevent the spread of antibiotic resistance', News Release, 7 November 2017, www.who.int/news-room/detail/07-11-2017-stop-using-antibiotics-in-healthy-animals-to-prevent-the-spread-of-antibiotic-resistance

Page 140 Antibiotics in China:

'Can China Kick its Animal Antibiotic Habit?' *The Guardian*, June 2019, www.theguardian.com/environment/2018/jun/19/can-china-kick-its-animal-antibiotic-habit

Peter Collignon and Andreas Voss, 'China, what antibiotics and what volumes are used in food production animals?', www.ncbi.nlm.nih.gov/pmc/articles/PMC4415312/#CR11

Page 142 The World Health Organization has classified macrolides as 'critically important to humans':

'WHO list of Critically Important Antimicrobials for Human Medicine', 5th revision, 2016, www.who.int/foodsafety/publications/cia2017.pdf?ua=1

Page 143 Tylosin and antibiotics:

According to the APVMA, 'In 2011, the WHO Advisory Group on Integrated Surveillance of Antimicrobial Resistance published the 3rd revision of their list of Critically Important Antimicrobials for Human

Medicine. Macrolides (including tylosin, oleandomycin and kitasamycin) were listed as critically important for human medicine.'

https://apvma.gov.au/sites/default/files/publication/27956-macrolides_prd_final.pdf

Page 143 2017 APVMA report:

Macrolide antibiotics (kitasamycin, oleandomycin and tylosin): Proposed regulatory decisions. The reconsideration of products containing kitasamycin, oleandomycin and tylosin and approvals of their associated labels, https://apvma.gov.au/node/27956

Page 144 Antibiotic resistance on big farms in the US:

www.foodandwaterwatch.org/sites/default/files/Antibiotic%20 Resistance%20101%20Report%20March%202015.pdf

Page 144 *National Geographic* story on methicillin-resistant *Staphylococcus aureus* (golden staph) in pigs and pig farmers:

www.nationalgeographic.com/science/phenomena/2009/01/23/mrsa-in-pigs-and-pig-farmers/

Page 145 Some data on MRSA in Swiss pigs:

www.sciencedirect.com/science/article/pii/S0378113515301103

Page 145 Data on the rapid increase in antibiotic resistance in Swiss pigs:

www.researchgate.net/publication/51247624_The_increase_of_methicillin-resistant_Staphylococcus_aureus_MRSA_and_the_presence_of_an_unusual_sequence_type_ST49_in_slaughter_pigs_in_Switzerland

www.aramis.admin.ch/Default.aspx?DocumentID=14339&Load=true

Confirming a 10-fold increase in four years: www.sciencedirect.com/science/article/pii/S0378113515301103

Page 145 MRSA in Dutch pigs:

https://academic.oup.com/jac/article/71/9/2414/2238495

Page 145 MRSA and its ability to cross species, and the new variants found in pig farms and on raw meat:

https://journals.plos.org/plosone/article?id=10.1371/journal.pone.0009385

Page 146 700 million pigs on farms in China:

www.reuters.com/article/us-china-pigs-production/china-supersizes-pig-farms-to-cut-costs-in-worlds-top-pork-market-idUSKBN1FQ39W

Page 147 Health impacts of animal-induced variants of MRSA hard to quantify:

http://umash.umn.edu/wp-content/uploads/2015/12/pigs-mrsa-112013.pdf

Page 147 Antibiotics and increase in chicken growth rates:

'Dietary antibiotics have been used in the food animal industry for more than 60 years, not only to control infectious diseases, but also to increase feed efficiency and improve growth performance. In chickens, subtherapeutic, in-feed antibiotics can increase body weight gain up to 8% and decrease the feed conversion ratio (feed intake/body weight gain) up to 5%, both compared with an antibiotic-free diet. However, use of antibiotic growth promoters in food animal production has led to the development of antibiotic resistance among the commensal gut microflora, thus increasing the zoonotic risk such as potential to be transferred to humans.'

www.nature.com/articles/s41598-018-22004-6

Page 147 'How Drug-Resistant Bacteria Travel from the Farm to Your Table', *Scientific American*, Dec 2016: www.scientificamerican.com/article/how-drug-resistant-bacteria-travel-from-the-farm-to-your-table/

Page 148 'Resistance Fighter', a profile on Stuart Levy, who promotes the idea of antibiotic-free livestock and is the past chairman and president of the Alliance for Prudent Use of Antibiotics (APUA):

The Scientist, June 2015, www.the-scientist.com/profile/resistance-fighter-35388

CHAPTER 14 Banned in 160 Nations – But Not Here

Page 158 Australian Pork Limited's Supplementary Submission to The Food Standards Amendment (Truth in Labelling Laws) Bill 2009 Inquiry:

A Response to the Humane Society Submission (submission number 12): Quote on pST (reporcin) from APL in a submission to the Australian Government, October 2009. (Downloads as a PDF from the Australian Parliament House website.)

The Interaction between Ractopamine Supplementation, Porcine Somatotropin and Moisture Infusion on Sensory Pork Quality, 2009. (Essentially how the chemicals used to make pigs bigger quicker make them less nice to eat, and how if you pump the pork full of water it helps to alleviate this problem):

https://pdfs.semanticscholar.org/5354/ea25e9290d001eb3bce53ea466f2b386b47f.pdf

Page 158 Improvac (increased pig weights and leanness):

http://slideplayer.com/slide/275868/

Page 160 Rivalea's stockfeed, Grolean:

www.rivalea.com.au/Stockfeed#418

Page 161 Grolean available in the US:

www.paybackfeeds.com/tag_book/documents/3535_
GLGROWER50_140T20.pdf

Page 162 16,000 dead pigs in Huangpu River, which leads to Shanghai:

www.theguardian.com/world/2013/mar/29/dead-pigs-china-water-supply

CHAPTER 15 What's Behind the Locked Gates?

Page 166 General chicken meat production from the industry peak body, the Australian Chicken Meat Federation:

www.chicken.org.au/chicken-meat-production/

Sheds can be 150 metres long and 15 metres wide (2250 square metres). This equates to a possible 18–20 birds per square metre, giving each bird the average space of an A4 piece of paper or similar.

Page 174 Baiada's fine for deceptive conduct:

The Australian, 8 July 2013, www.theaustralian.com.au/news/nation/
poultry-producer-baiada-misled-over-free-to-roam-chickens-accc-finds/
news-story/a518ebbd7396cc4df68b41ae1536df47

www.accc.gov.au/media-release/court-orders-chicken-companies-to-pay-
400000-for-'free-to-roam'-misleading-claims

CHAPTER 16 How Do We Feed the World? Animals May Well Be Part of the Plan

Eric Garza teaches at the University of Vermont and specialises in energy use in food systems:

https://learn.uvm.edu/foodsystemsblog/2014/07/10/meat-vs-veg-an-energy-
perspective/

Page 179 World population, past, present and future (10 billion by 2055):

www.worldometers.info/world-population/

UN population projections (9.8 billion by 2050):

www.un.org/development/desa/publications/world-population-prospects-
the-2017-revision.html

Page 180 Food and Agriculture Organization research, published in Global Food Security, on biomass not edible by humans, but by livestock:

www.fao.org/ag/againfo/home/en/news_archive/2017_More_Fuel_for_the_Food_Feed.html

Page 183 David Pimentel estimates that if all humanity ate the way Americans eat, we would exhaust all known fossil fuel reserves in just seven years:

www.ecoliteracy.org/article/fossil-food-consuming-our-future#

Page 187 You can feed more people, in the US (and potentially globally) using an omnivorous diet, better than vegan, on available land:

www.elementascience.org/articles/10.12952/journal.elementa.000116/

Page 188 An article on micronutrients:

www.cdc.gov/immpact/micronutrients/

Page 189 70–80% of grain grown in US is fed to livestock:

www.ecoliteracy.org/article/fossil-food-consuming-our-future#

Page 189 The average foodstuff on an American supermarket shelf has travelled 1500 miles:

www.ecoliteracy.org/article/fossil-food-consuming-our-future#

Page 189 'All told, it takes 68 calories of fossil fuel to produce one calorie of pork, and 35 calories of fuel to make one calorie of beef':

www.ecoliteracy.org/article/fossil-food-consuming-our-future#

Page 189 84% of vegetarians and vegans return to an omnivorous diet:

www.psychologytoday.com/au/blog/animals-and-us/201412/84-vegetarians-and-vegans-return-meat-why

The original report is summarised by the people who funded it, Faunalytics, here:

https://faunalytics.org/how-many-former-vegetarians-and-vegans-are-there/

Page 190 Australian agriculture contributes about 13% of its total greenhouse gas emissions, data from the Climate Council:

www.climatecouncil.org.au/from-farm-to-plate-to-the-atmosphere-reducing-your-food-related-emissions/

Page 190 We already grow enough food, internationally, to feed one and a half times the world population (over 11 billion people):

Food and Agriculture Organization of the United Nations: www.researchgate.net/publication/241746569_We_Already_Grow_Enough_Food_for_10_Billion_People_and_Still_Can't_End_Hunger

Huffington Post, 2012, www.huffpost.com/entry/world-hunger_n_1463429

Page 190 Some good research on how to feed the increasing world population through system changes:

www.elementascience.org/articles/10.1525/elementa.310/

CHAPTER 17 The No-Meat Diet Isn't the Answer. Neither Is the Attitude

Page 194 Good US data on veganism and vegetarianism:

https://faunalytics.org/how-many-former-vegetarians-and-vegans-are-there/

Page 195 *A Pound of Flesh; A survey of 1202 Australians about whether they're vegetarian or vegan and what their attitudes to animals are*, Newspoll report commissioned for the Vegetarian/Vegan Society of Queensland, 2010. A copy can be downloaded at www.voiceless.org.au

Vegetarian Victoria bemoaning the lack of data by the Australia Bureau of Statistics on vegetarianism: https://vegvic.org.au/statistics/

Page 196 The top 20 sources of vitamin A, and some vegetables that can provide the nutrient (though not all people have the genetic ability to absorb all the vitamin A available through the body's processing of carotenoids into vitamin A):

www.healthline.com/nutrition/foods-high-in-vitamin-a#section2

Page 196 How to get enough vitamin A in your diet as a vegan:

www.theveganrd.com/2009/04/fruits-and-veggies-vitamin-a-and-vegan-diets/

Page 198 *Guardian* Podcast, *We Need To Talk About: Veganism*, 2018:

www.theguardian.com/membership/audio/2018/may/30/we-need-to-talk-aboutveganism

Page 200 US Food & Drug Administration's *Defect Levels Handbook*:

www.fda.gov/Food/GuidanceRegulation/
GuidanceDocumentsRegulatoryInformation/SanitationTransportation/
ucm056174.htm#CHPTA

Page 200 5% of the weight of hops can be aphids:

Layla Eplett, 'I hate to break it to you, but you already eat bugs', *Scientific American*, June 2013, https://blogs.scientificamerican.com/but-not-simpler/i-hate-to-break-it-to-you-but-you-already-eat-bugs/

CHAPTER 18 Eat Dairy? You'd Better Eat Veal, Too

Page 203 RSPCA view on bobby calves (male calves in the dairy industry):

https://kb.rspca.org.au/what-happens-to-bobby-calves_87.html

Page 205 RSPCA view on 'quick maceration' – the killing of day-old male chicks from laying hens:

https://kb.rspca.org.au/what-happens-with-male-chicks-in-the-egg-industry_100.html

Page 207 By the late 19th century, 159,000 ducks were passing through the Melbourne markets a year:

http://theconversation.com/ducking-and-weaving-should-we-be-hunting-in-australias-wetlands-148

CHAPTER 19 Fake Meat

Page 210 Australia's food footprint:

www.foodchoices.com.au/samples/resource%20sheet%2056.pdf

Page 210 US data on agriculture, including land use, water, emissions:

https://academic.oup.com/ajcn/article/78/3/660S/4690010

Page 210 Stats on farming land vs calories:

www.theguardian.com/environment/2018/may/31/avoiding-meat-and-dairy-is-single-biggest-way-to-reduce-your-impact-on-earth

Page 211 Professor Michael Crawford and Yoqun Wang of London Metropolitan University found that a chicken in 2004 contained more than twice as much fat as in 1940, a third more calories and a third less protein:

'It's supposed to be lean cuisine. So why is this chicken fatter than it looks?', *The Guardian*, May 2005, www.theguardian.com/lifeandstyle/2005/may/15/foodanddrink.shopping3

Page 212 The drop in price of lab-grown meat:

'The future of meat is fake – grown in a laboratory from animal cells', *The New Daily*, May 2018, https://thenewdaily.com.au/life/wellbeing/2018/05/05/fake-meat-laboratories/

Page 213 To produce so-called 'clean meat', cultured meat, an animal still has to die. And it has to have flavours and colours added:

www.abc.net.au/news/2018-05-06/vegan-alternative-plant-based-meat-grown-in-lab/9726436 www.theguardian.com/environment/2018/apr/30/lab-grown-meat-how-a-bunch-of-geeks-scared-the-meat-industry

Page 216 Great story on lab meat and the energy it takes to produce:

www.abc.net.au/triplej/programs/hack/chicken-nuggets-lab-grown-from-feathers-to-go-on-sale-soon/10397808

Page 216 ABC story on lab-grown meat, which confirms price drop:

www.abc.net.au/news/rural/2018-05-06/vegan-alternative-plant-based-meat-grown-in-lab/9726436

Page 217 ABC story on who might eat lab-grown meat:

www.abc.net.au/news/science/2017-12-01/lab-made-meat-can-food-be-more-humane-and-less-polluting/9206542

Page 217 Customer perceptions of lab-grown meat:

https://theconversation.com/would-you-eat-meat-from-a-lab-consumers-arent-necessarily-sold-on-cultured-meat-100933

Page 217 Lab-grown meat and fish:

www.theguardian.com/lifeandstyle/2017/sep/20/lab-grown-meat-fish-feed-the-world-frankenmeat-startups

Page 217 The trouble with fake meat, *The Guardian*, January 2019:

www.theguardian.com/food/2019/jan/27/the-trouble-with-fake-meat-beetroot-burgers-food-substitutes

Page 219 70% of the world's food is grown by smallholders:

www.huffingtonpost.com/hugh-locke/smallholder-farmers-are-t_b_7865848.html

Page 219 Food and Agriculture Organization information that small farms produce 80% of the world's food:

www.fao.org/news/story/en/item/260535/icode/

'Globally, 60–80% of peasant growers are women; in Africa, women produce 90% of the continent's food supply and contribute to about half of global food supply':

Coping With the Food and Agriculture Challenge: Smallholders' Agenda, United Nations Conference on Sustainable Development, 2012, www.fao.org/fileadmin/templates/nr/sustainability_pathways/docs/Coping_with_food_and_agriculture_challenge__Smallholder_s_agenda_Final.pdf

CHAPTER 20 **When the Hunter Becomes the Hunted**

Page 221 Australian Bureau of Statistics on the number of people who fish recreationally:

www.agriculture.gov.au/abares/Pages/recreational-and-charter-fishing.aspx

Page 222 Meet a duck hunter:

www.abc.net.au/news/2018-04-11/duck-hunting-on-lake-connewarre/9591250

Page 223 'Overseas research generally supports the contention that young birds are more naive and more likely to be shot than experienced adults':

'Ducking and weaving: should we be hunting in Australia's wetlands?' *The Conversation*, March 2011, https://theconversation.com/ducking-and-weaving-should-we-be-hunting-in-australias-wetlands-148

Page 224 ACCC on Luv-a-Duck, November 2013:

www.accc.gov.au/media-release/court-orders-luv-a-duck-to-pay-360000-for-misleading-claims

Page 224 In 2010 Luv-a-Duck were processing 100,000 birds a week:

www.thecourier.com.au/story/5723802/luv-a-duck-toasts-50-years-of-operation-photos/

Page 224 The duck industry processes about 8 millions birds a year:

www.agrifutures.com.au/farm-diversity/ducks/

Page 224 Some 'ag gag' stuff about farmed ducks:

https://newmatilda.com/2018/06/11/ducking-hell-shocking-video-revealing-farm-ducks-australia/

Page 226 Feral camel cull:

www.news.com.au/technology/science/animals/australian-feral-camel-population-estimated-to-be-at-12-million-and-growing-fast/news-story/aa162cbe2d1c0b2fc306f6c0aadde341

www.environment.gov.au/system/files/resources/2060c7a8-088f-415d-94c8-5d0d657614e8/files/feral-camel-action-plan.pdf

Page 226 National Feral Camel Action Plan: www.environment.gov.au/system/files/resources/2060c7a8-088f-415d-94c8-5d0d657614e8/files/feral-camel-action-plan.pdf

136,000 camels shot under the plan in about five years: www.publish.csiro.au/rj/RJ15100

Page 226 5 million plus kangaroos shot in 2012:

www.bbc.com/news/world-australia-35878526

www.nationalgeographic.com.au/australia/australians-hunt-kangaroos-commercially-does-it-make-sense.aspx

Page 226 Feral horse, pig, buffalo and donkey cull in Kakadu:

Michael Murdoch, ABC Rural, 17 March 2017, 'Thousands of wild horses, donkeys, buffalo and pigs shot in Kakadu cull', www.abc.net.au/news/rural/2017-03-17/kakadu-feral-animal-cull-buffalo-donkeys-pigs-horses/8363580

Page 226 Feral pig cull in Queensland:

www.news.com.au/national/queensland/marksmen-in-helicopters-cull-more-than-30000-feral-pigs-in-north-queensland-as-rare-turtles-face-extinction/news-story/84ef26393ae99b43ffcd437f38c356d0

Page 226 'Wild boar exports suffer as accredited hunter numbers drop': ABC News, April 2015, www.abc.net.au/news/rural/2015-04-14/wild-boar-market-hit-by-dropping-hunter-numbers/6383368

Page 227 King Island wallaby cull:

www.bbc.com/news/world-australia-35878526

Page 227 Tasmanian wallaby cull:

www.themercury.com.au/news/tasmania/with-a-population-explosion-under-way-is-wallaby-meat-tasmanias-most-wasted-resource/news-story/e369258ebbb221f27fa6d2dbfe9d0136

Page 227 Over a million wallabies and 400,000 possums shot for crop protection in 2006/7 still not considered effective by farmers:

https://dpipwe.tas.gov.au/Documents/1080ReportFINAL.pdf

CHAPTER 25 The True Cost of Food

Page 262 Average food expenditure dropped as a proportion of income; 'household food expenditure as a share of after-tax income fell from 19% in 1988–89 to 14% in 2015–16' – that's after a massive increase in takeaway and dining out:

Department of Agriculture and Water Resources, http://data.daff.gov.au/data/warehouse/9aat/2018/fdati9aat20180822/FoodDemandInAustralia_20180822_v1.0.0.pdf

Page 263 Some historical data on how meat has dropped in cost:

www.sciencedirect.com/science/article/pii/S0264999315000656

In a 2011 Murdoch University report, Australians spent about 4% of their weekly budget on meat – down from about 8% in 1962 – while meat consumption has increased from about 100 kg to about 110 kg per year: www.murdoch.edu.au/School-of-Business-and-Governance/_document/ Australian-Conference-of-Economists/Changing-pattern-of-meat-consumption-in-Australia.pdf

Page 263 Cost of owning a pet:

www.moneysmart.gov.au/life-events-and-you/life-events/getting-a-pet/ the-cost-of-a-pet

Page 267 Global meat consumption, past and present:

https://ensia.com/articles/these-maps-show-changes-in-global-meat-consumption-by-2024-heres-why-that-matters/

Page 268 Indian antibiotic use in livestock:

www.theguardian.com/environment/2018/feb/01/indias-farmed-chickens-dosed-with-worlds-strongest-antibiotics-study-finds

Tom Batchelor, 'Farmers using powerful antibiotic "of last resort" on healthy chickens to boost weight, report finds', *The Independent*, February 2018, www.independent.co.uk/environment/farmers-chickens-antibiotic-boost-weight-colistin-farm-animals-infections-a8191521.html

Page 268 200 million eggs recalled due to a salmonella scare:

www.nytimes.com/2018/04/15/business/egg-recall-salmonella.html?partner =rss&emc=rss&smtyp=cur&smid=tw-nytimes

Salmonella appeared, in several countries at about the same time, in eggs (hence chooks) in the late 1970s early 1980s: David Brown, 'Egg-loving salmonella bacteria have been sickening people for decades', *Washington Post*, 14 September 2010, www.washingtonpost.com/wp-dyn/content/ article/2010/09/13/AR2010091303594.html

INDEX